The State of Butterflies in Jersey

- The Jersey Butterfly Monitoring Scheme: 2004 to 2013 -

STATES OF JERSEY
2015

The State of Butterflies in Jersey

First published in Great Britain
in 2015 by Société Jersiaise
www.societe-jersiaise.org

A catalogue record of this report is available from the British Library

ISBN 978 0 901897 55 8

Contents

In dedication to
MARGARET LONG and JOAN BANKS

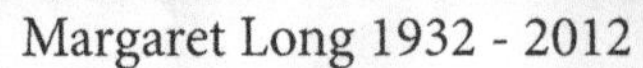
Margaret Long 1932 - 2012

Joan Banks 1937 - 2011

Margaret and Joan were instrumental in setting up the Jersey Butterfly Monitoring Scheme. They had tested a similar scheme in the 1990s and in 2004 used the data from this to establish butterfly transects around the island. They also provided background knowledge on Jersey's butterflies and continued to support the scheme for many years. Without them the Jersey Butterfly Monitoring Scheme would have been much harder to get off the ground. We remain very grateful for all their help and support.

Foreword

It is an honour and a pleasure to provide a foreword for this superb study of Jersey's splendid butterflies, and valuable account of the, at least recent, history of their status, as well as a somewhat worrying look at what the future holds for them. Our butterflies' problems are largely similar to those affecting British and other populations, and it is only with information such as is now available in this report that action may be attempted to halt declines, and to restore butterflies to their once familiar roles in countryside and garden.

My pleasure is further enhanced by seeing our 'work' - if butterfly-watching can be called work - of very many years ago making useful contributions to present-day studies. Of course others enjoyed seeing and noting them in those distant days but, fifty years ago, butterflies were far more taken for granted with little thought given to possible extinction, and counting and recording were restricted to the keener naturalists, who could be seen as unwittingly following one eminent scientist's rather broad claim that 'science is counting'. Here I must emphatically dissociate myself from any credit attaching to these early steps in recording the numbers and status of butterflies in Jersey. The book's introduction has a full account of the early workers and publications listing species to be found in the island, but few then kept records of numbers. I know that one who did was my late wife, Margaret Long, determinedly looking for butterflies throughout all their seasons, and I, not unwillingly, was drawn in, watching and helping her embark on this approach to natural history, sketchily at first but soon increasingly, with her eventually sharing its demands with her beloved botanical studies. Thus I bask in reflected glory.

Gradually members of other natural history sections of the Société Jersiaise added their observations, and valuable contributions from the fieldwork of the late Joan Banks, and the fine photography of Richard Perchard provided considerable support for longer-term recorders.

Margaret was constantly consulted by most of the many graduates in the Ecology Department of University College, London, who spent nearly twenty summers in the island studying for higher degrees, and she assisted particularly Hannah Clarke whose recommendations for monitoring may be seen as the genesis of much that follows herein.

As well as walking their own transects Margaret and Joan assisted with reinstating some transects that had been previously monitored and created new ones.

Credit is equally due to the scheme's several dozen recorders, conscientiously tramping their transects, accumulating ten years' worth of extremely valuable data, even walking less productive sites, sustained by the knowledge that they may be contributing to ensuring the conservation of this major component of Jersey's wonderful natural history.

Congratulations are due to the Department of the Environment Natural Environment Team for their initiative and continued support for the scheme, and especially to Paul Chambers for master-minding the analyses of the data to produce this instructive and convincing, if a little disturbing, book.

ROGER LONG
MAY 2015

The Importance of Monitoring Butterflies

At first glance the idea that the States of Jersey should be encouraging people to count butterflies might look unusual and perhaps even quaint. In reality Jersey's butterfly monitoring scheme is a cutting edge scientific project whose results perform a vital role in terms of environmental understanding at a domestic level but also as part of a wider European network.

It is sometimes hard to communicate the importance of projects such as this and so this introductory preamble will attempt to answer some of the questions that we are commonly asked about the Jersey Butterfly Monitoring Scheme.

Why does Jersey monitor butterflies?

In 1992 Jersey signed the International Convention on Biological Diversity (CBD) at the Rio Earth Summit. In 2000, the States of Jersey published its *Biodiversity Strategy* which recognised that fulfilment of Article 7 of the CBD necessitated establishing robust programmes for monitoring the health of indicator species and key habitats.

The effective and economical monitoring of environmental conditions is best achieved by using single issues or events as indicators. After wide consultation it was decided that monitoring a key biological population (such as butterflies) by looking at the species' distribution, abundance and changes over time would provide robust statistical information about the state of Jersey's environment.

Using butterflies as an indicator was a rational choice. In the UK butterflies have been used as key environmental indicators since 1976 and by 2000 the British-developed methodology had become a standard monitoring tool for governments across the European Union. With a network of butterfly monitoring projects already in operation across Europe, it was a logical step for Jersey to adopt a methodology that was demonstrably cost effective, scientifically robust and whose results were comparable with those from other countries.

The Jersey Butterfly Monitoring Scheme (JBMS) was established in 2004 as part of the Department of the Environment's integrated 'State of the Environment' monitoring programme.

What makes butterflies so useful as environmental indicators?

Butterfly monitoring is not undertaken because these animals are pretty but for sound scientific reasons. Butterflies are insects, a group which represents over half of all terrestrial biodiversity. Unlike most insects, butterflies are large, easy to identify, popular with amateur naturalists and do not bite, sting or pinch. They have a complex lifecycle that starts with an egg which will hatch into a herbivorous caterpillar. The caterpillar will then feed and pupate into the familiar airborne adult form. The adults (which are pollinators) then mate and lay further eggs.

Every stage of a butterfly's lifecycle is dependent upon individual plant species which will grow in particular habitats or ecological niches. If the habitats and niches are altered or removed then there will be an immediate effect on the local butterfly population. A sustainable butterfly population requires a network of breeding and feeding habitats scattered across the landscape. Any changes to this network (good or bad) can be immediately picked up through monitoring; it is this fine scale reaction to change that makes butterflies such excellent environmental indicators.

Who monitors Jersey's Butterflies?

The JBMS is wholly reliant on members of the public volunteering to walk a set route once a week between April and September. The volunteers identify and count butterflies under strict conditions and return their data to the Department of the Environment. This is a great example of citizen science and, thanks to the dedication of the volunteers, is exceptionally cost effective.

Like many of Jersey's monitoring schemes, the butterfly scheme is dependent on the good will and talent of the island's public. Without their dedication the island would probably not be able to meet its commitments to the CBD and other international environmental agreements.

What happens to Jersey's butterfly data?

The JBMS raw data are collected and collated by the Department of the Environment with copies being passed on to the Jersey Biodiversity Centre, Société Jersiaise, the UK Butterfly Monitoring Scheme, Butterfly Conservation and the EU Environment Agency. The island has 40 monitored sites (which is more than several European countries) and punches well above its weight. JBMS data are used in local, national and international analyses such as the European Grassland Indicator Butterfly Scheme. The results are analysed annually and after 10 years of continuous monitoring the data obtained enough statistical significance to undergo a more thorough analysis, the results of which are presented in this book.

What does the JBMS tell us about the island's environment?

The JBMS 10 year results suggest that Jersey's butterflies respond quickly to changes in the environment so are thus an excellent indicator of changes in the island's terrestrial habitats and climate. The results suggest that there has been an overall decline in Jersey's butterflies since 2004, especially on agricultural and urban sites, but that managed semi-natural sites are mostly doing well. Now that these and other issues have been highlighted by the JBMS, it may be possible to help mitigate and reverse any declines in species and habitat quality through government policy and changes in land management practice.

- Part One -

The
Jersey Butterfly Monitoring Scheme

1.1 - The JBMS: Background

A Brief History of Butterfly Studies in Jersey

Although butterflies have been studied and admired in Britain for centuries, the first list of Jersey species did not appear until 1862. It was compiled by F.G. Picquet for a general guidebook to the Channel Islands and, while comprehensive, has some entries that have since been questioned (Ansted and Latham, 1862; Picquet, 1873; see also Section 2.2).

In 1873 the foundation of the Société Jersiaise provided a framework within which historians, naturalists, linguists and other learned people could collectively undertake and publish their research. Almost from the outset the Société Jersiaise had a Natural History Committee which, as time progressed, became subdivided into different specialist sections covering plants, fungi, insects, birds, marine biology, etc.

By the start of World War One an Entomology Section had been formed and was actively gathering specimens from all Jersey's butterfly species (Luff, 1908). In 1916 this collection stood at 497 specimens from 33 species and by 1918 the Section had taken so many specimens that almost all its drawer space had been used up (Clarke, 1991).

Local entomological studies experienced a lull during the 1920s and in this time the island received an occasional visit from UK naturalists (e.g. Riley, 1922). From 1930 onwards the Entomology Section was reinvigorated by A.C. Halliwell who, in 1932, produced a list of 38 Jersey butterfly species, of which three were thought to be extinct and six to be rare migrants. This work was furthered by Walter Le Quesne who took over the Entomology Section in 1946 and was responsible for records of a number of rare species (most notably the Large Chequered Skipper) during and immediately after the German Occupation (Le Quesne, 1946).

Some years after World War Two the Société Jersiaise became home to a group of talented naturalists which included Roger and Margaret Long, Frances Le Sueur, Joan and John Banks, John Richards and others. Their work was to transform our knowledge of local natural history and many of their studies underpin Jersey's present day environmental and ecological monitoring, including the JBMS. In 1970 Roger Long published a revised list of Jersey butterflies which was summarised and updated a few years later in Frances Le Sueur's *Natural History of Jersey* (Long, 1970; Le Sueur, 1976).

A pioneering attempt at systematic monitoring of local butterflies was undertaken by Hannah Clarke during the summer of 1991. Clarke established seven transects across Jersey and monitored them for two months with the same methodology as used by the JBMS (see Section 1.2). Clarke recorded 28 resident and regular visitor species and one rare migrant (Queen of Spain Fritillary). This study contains some of the last local records for the Large Chequered Skipper and expressed concerns about the health of Jersey's butterflies. It was recommended that Jersey should adopt a wider monitoring scheme (Clarke, 1991).

General entomological studies continued at the Société Jersiaise and from 1985 onwards the Department of the Environment (States of Jersey) commissioned a series of wildlife surveys (which included invertebrates) at specific sites. Between 2001 and 2004 a government audit of local biological research and wider environmental monitoring led to the establishment of several projects which could provide benchmark data to monitor progress in key areas of the local environment. This included the Jersey Butterfly Monitoring Scheme (JBMS) which was established in 2004 under the management of Nina Cornish at the Department of the Environment (States of Jersey, 2005; 2011). The JBMS was a long term project which required ten years of data before its results could be deemed statistically significant (see below).

The JBMS continued to operate successfully and in the meantime revisions of local butterfly lists were issued by Schaeffer (2008) and Long (2009). These suggest that the Glanville Fritillary and Large Chequered Skipper became locally extinct following Clarke's 1991 survey.

In 2013 the JBMS achieved its full ten-year dataset. This report is a summary of the analytical results of that dataset and represents the most in-depth study into Jersey's butterflies. However, that the JBMS could have existed at all is because it was able to build upon a tradition of local butterfly studies that stretches back over a century.

In 2012 the need for a central repository for all the Island's biological records led to the establishment of the Jersey Biodiversity Centre. This has greatly assisted local wildlife studies, including lepidoptera, and provided a publicly accessible repository for the JBMS's records.

The Jersey Butterfly Monitoring Scheme

Systematic wildlife monitoring is an invaluable tool for nature conservation as it can identify trends and changes within local and regional environments that will not be obvious from casual recording alone. In northern Europe butterflies are an ideal subject for wildlife monitoring as they are large, day-flying animals that may be found across a wide range of habitats. Butterflies are also common, easy to identify and well-known to (and admired by) many amateur and professional naturalists.

Butterfly monitoring in the UK started in the 1960s and has continued via several schemes. Results from these schemes suggest that butterflies are good indicators of general environmental health and that the long-term study of their abundance, diversity and distribution can be used to detect fine-scale changes in habitat, environment, climate and wider environmental biodiversity.

Several biological factors make butterflies sensitive to environmental change including a short life-cycle, a reliance on specific larval foodplants, a sensitivity to climate conditions and an inability for populations to disperse themselves rapidly. Also, many butterfly species are restricted to specific habitats and being herbivorous insects places them low down on the food chain. It is for these reasons that butterflies are being used as headline indicators for wider environmental health and climate change by several governments (including the UK) and the European Union. (For further information see Thomas, 2005; Parmesan, 2003; Fox *et al.*, 2006, van Swaay *et al.*, 2008.)

The national UK Butterfly Monitoring Scheme (UKBMS) has been running since 1976 and has proved to be a successful and effective tool for monitoring the changes in ecology and climate. When the Jersey Butterfly Monitoring Scheme (JBMS) was founded in 2004 it was decided to adopt the same methodology as the UKBMS.

The JBMS wanted to ensure there was good coverage of all the island's habitats and so rather than let its volunteers choose their own sites to monitor (as the UKBMS has done, leading to a concentration within natural areas), the Department of the Environment specified 28 monitoring sites which covered wildlife, agricultural and urban areas.

During its first year the JBMS recruited seventeen volunteers to make weekly butterfly counts during the spring and summer. Many of these original volunteers are still with the JBMS and the time and effort that they and more recent recruits have devoted to the scheme cannot be overestimated. An insight into just how much time and energy has been devoted to the JBMS can be seen in the statistics presented in Table 1.

The JBMS has been managed by the Department of the Environment and 2013 saw the completion of its tenth continuous year of monitoring. During 2014 the ten-year dataset was checked for errors and then analysed with a view to determining the state of health of Jersey's butterflies. This analysis and the presentation of its results follow methodologies and statistical techniques that were devised by the UKBMS for their reports.

Since its inception the JBMS has grown considerably and a further ten transects have been created to accommodate new volunteers to the project. By 2013 the JBMS had received help from over 50 volunteers who between them have monitored a total of 38 transects covering Jersey's main habitat types. Every year the JBMS organises a local butterfly meeting and training event, complete with guest speakers. This allows everyone to see the previous year's results and encourages new volunteers.

As well as the ten-year analysis, the JBMS data form part of the five-yearly 'State of Jersey' environmental assessment (States of Jersey, 2011). Its results are shared with the Centre for Ecology and Hydrology, the Société Jersiaise and the Jersey Biodiversity Centre.

By any measure the JBMS has been a great success and, according to Butterfly Conservation, the resultant ten-year dataset is excellent. The JBMS continues to operate and, to judge by the patience and goodwill of its volunteers, will it is hoped be going for many years to come. The scheme has produced an extraordinary amount of robust data, the analysis of which provides a snapshot into the state of health within Jersey's butterfly populations.

None of this would have been possible without the JBMS's army of volunteers. That it has been possible to produce a detailed report such as this, is entirely down to their dedication and hard work. We cannot thank them enough for the hundreds of hours they have spent counting thousands of butterflies. This report is therefore dedicated to each and every person that walked our transects, clipboard in hand, hoping for the glimpse of a butterfly as it flutters all-too-briefly across our path. Thank you, one and all.

JBMS Statistics: 2004-2013

Number of transects	38
Shortest transect	84 metres
Longest transect	3.5 km
Average transect length	991 metres
Total no. of individual sections	227
Total distance walked	5,904 km/3,668 miles
Total no. of walks	5,606
Total no. of volunteers	52
No. of butterflies counted	122,279
No. of species encountered	34

Table 1. A selection of statistics illustrating the scope and scale of the JBMS and the amount of time and effort devoted to it by its volunteers.

1.2 - The JBMS: Methodology

The JBMS uses a transect-based methodology developed by Dr Ernie Pollard in the 1970s and then pioneered by the UK Butterfly Monitoring Scheme (UKBMS). This methodology was chosen for Jersey because it is simple, statistically robust and has a proven record of success with the UKBMS. It also fulfils the environmental monitoring requirements of the JBMS when it was founded.

The JBMS monitoring scheme is based on a series of transects, each of which is a fixed route across the landscape that is walked weekly by a volunteer. Most JBMS transects are subdivided into several sections, each of which represents a particular habitat or style of land management (e.g. a grassy field, area of scrub, sand dune, etc.). The JBMS transects vary from just 84 metres in length to over 3.5 kilometres and have between one and fourteen sections (Table 1).

Between 1st April and 30th September each year a volunteer will walk their allotted transect once a week. There are a total of 26 butterfly weeks each year and, as with every other aspect of the JBMS methodology, the Jersey monitoring calender matches the one used by the UKBMS.

Transect recording must take place in weather conditions that are suitable for butterfly flight. In practice this means that the walk should take place between 10 am and 5 pm (but ideally between 10.45 am and 3.45 pm). If the air temperature is above 17°C then the walk should be in at least 40% sunshine or, if the temperature is above 13°C, 60% sunshine. An individual walk is considered to be invalid if it took place in the rain, if the temperature was below 13°C or if the wind was above Force 5.

As the volunteer walks along the transect route, they should identify and count any butterflies that fly within an imagined box that stretches 2.5 metres either side of them (or 5 metres to one side if walking next to a hedge or bank), 5 metres ahead of them and 5 metres above the ground (Fig. 1).

Any butterflies that are observed in the distance but do not enter this box are ignored as are those whose identification cannot be established for certain. Butterfly counts are recorded on special forms together with notes about the weather, habitat, land use and other observed wildlife such as day-flying moths, etc.

For a more detailed explanation of this methodology see Pollard and Yates (1993), Fox *et al.* (2006), the UKBMS website (www.ukbms.org) or contact the States of Jersey (Department of the Environment).

Data Collection, Processing and Analysis

At the end of September each year JBMS volunteers return their completed count forms to the Department of the Environment. Their species' counts, weather and habitat information are then inputted into Transect Walker, a bespoke piece of software developed by Butterfly Conservation to collate and analyse UKBMS data.

In 2014 preparation began for the JBMS dataset's ten-year analysis. All records from 2004 to 2013 were checked for inputting errors, missing data, unusual reports and other problems. This highlighted a relatively small number of issues that were addressed by checking the computer records against the original recording sheets. A copy of the data was then moved into an Access database where it could be manipulated and queried more easily than in Transect Walker.

Since 2004 each section of each transect has been given habitat and land management classifications according to the European EUNIS scheme. Any changes in a section's habitat or management over a season or between years would be noted by the volunteer and then recorded in Transect Walker. This allows the matching of individual butterfly counts to the habitat and land use classification of the sections where they were recorded.

In 2011 the whole of Jersey was surveyed and mapped using the JNCC's Phase 1 habitat classification. This survey provided detailed habitat information which could be utilised by the JBMS. Therefore in 2014 the JBMS section habitats, which were initially classified using the EUNIS scheme, were reclassified into the Phase 1 scheme. In this report it is the Phase 1 scheme that is used when habitat classification information is provided.

Figure 1. (Left) An aerial photograph and map showing a butterfly transect route in St Helier. The sections are marked out in blue and red. (Right) When walking a transect the volunteer must only record butterflies that come within an imaginary box with 5 metre sides.

During 2014 the corrected ten-year JBMS dataset was analysed using a built in statistical function within Transect Walker. This generated an yearly index figure which represents the annual count for each butterfly species on each transect after taking into account any missed weeks.

The annual index figures produce by Transect Walker were then analysed using the software TRIM which uses a log-linear Poisson regression model to generate an overall index number for each species for each year of monitoring. These annual indices can be used to calculate how the overall abundance of each species has changed annually between 2004 and 2013 (see Section 3.1).

As well as being statistically robust, these indices of abundance, when plotted by year, provide information on the ten-year trend within individual butterfly populations (See Section 2.1). For more information on these statistical techniques see Fox *et al.* (2006). As as well using indices of abundance, other statistical techniques were also applied to the JBMS data such as calculating the average number of butterflies sighted per kilometre walked.

The ten-year results paint a mixed picture regarding the state of Jersey's butterflies and highlight some areas of concern. It is hoped that some of the issues outlined in this report can be addressed and that the measured decline in some of the island's butterfly population can be slowed or even reversed. It is probable that the next major analysis of JBMS will occur when the JBMS has fifteen or twenty continuous years of monitoring data.

The JBMS Transects

Since monitoring began in 2004, the number of transects operating within the JBMS has changed. The scheme began with 28 transects but as new volunteers joined, so the number of transects was expanded to accommodate them and, at the time of writing, there have been a total of 41 transects operating at one time or another. Data from 38 transects have been included in this analysis.

However, not every transect was monitored every year. There have been years when individual volunteers have had to drop out through illness, other commitments or because they simply could not find the time to undertake a walk each week. There are also three transects that have ceased altogether because their sites have been built on or developed in such as way that access has become restricted.

Fortunately the analytical technique developed by the UKBMS is designed to take into account missed weeks or even entire years. This, combined with our volunteers' diligence in filling their forms, meant there were few problems with the data submitted to the JBMS. The Jersey dataset and the results of its analysis have been seen by statisticians at Butterfly

No.	Transect Name	No. Sections	Year Started	Length (m)	Habitat Type
1	Fern Valley	4	2004	785	Semi-natural
2	Les Landes	14	2004	2,435	Semi-natural
3	St Catherine's Wood	8	2004	2,541	Semi-natural
4	Les Blanches Banques	8	2004	3,189	Semi-natural
5	Field A (St Ouen)	4	2004	621	Agricultural
6	Field B (St Mary)	11	2004	2,109	Agricultural
7	Field C (Trinity)	7	2004	1,261	Agricultural
8	Field D (Trinity)	3	2004	251	Agricultural
9	Field E (Trinity)	2	2004	261	Agricultural
10	Field F (Trinity)	2	2004	257	Agricultural
11	Field G (St Brelade)	6	2004	686	Agricultural
12	Field H (St Brelade)	7	2004	876	Agricultural
13	Field I (St Martin)	1	2004	84	Agricultural
14	Field J (St Lawrence)	10	2004	500	Agricultural
15	Field K (St Lawrence)	4	2004	330	Agricultural
16	Field L (St Brelade)	5	2004	559	Agricultural
17	Field M (St Brelade)	3	2004	234	Agricultural
18	Field N (St Ouen)	2	2004	476	Agricultural
19	Field O (St Ouen)	2	2004	678	Agricultural
20	Field P (Trinity)	1	2004	492	Agricultural
21	Field Q (Trinity)	3	2004	482	Agricultural
22	Grouville Golf Course	10	2004	3,569	Semi-natural
23	Les Mielles	6	2004	805	Semi-natural
24	La Commune	6	2004	686	Semi-natural
25	West Park	7	2004	988	Urban
26	Swiss Valley	9	2004	1,231	Semi-natural
27	Green Street Cemetery	6	2004	494	Urban
28	Howard Davis Park	10	2008	874	Urban
29	St Ouen's Pond	7	2005	1,098	Semi-natural
30	L'Oeillère Headland	12	2007	1,096	Semi-natural
31	Grainville School	6	2007	384	Urban
32	Victoria Tower	6	2008	442	Semi-natural
33	South Hill Park	5	2010	529	Urban
34	St John's Monument	8	2010	979	Semi-natural
35	La Sarsonnerie	9	2011	725	Semi-natural
36	Faldouet	5	2013	687	Semi-natural
37	Noirmont	7	2013	1,974	Semi-natural
38	Upper Dunes	7	2013	1,982	Semi-natural

Table 2. The 38 JBMS transects operating between 2004 and 2013.

Conservation and have been declared robust and statistically meaningful (see Section 3.1).

The JBMS transects were initially chosen to represent a range of habitats across semi-natural, agricultural and urban areas in Jersey. (See Section 3.2 for a definition of these terms.) As new transects have been added, so the range of monitored habitats has grown.

The JBMS transects are not evenly distributed around the island. There are more transects located around the edge of the island than in the middle. There are also more transects on the west of the island than the east. This is because there are more transects on semi-natural sites than there are ones on agricultural or urban ones and these tend to be sited near to the coast and in the west of Jersey.

However, the JBMS more than adequately covers Jersey's principal habitats and the density of transects is high when compared with the UKBMS. Table 2 provides some basic information on the 38 transects used in this study while Figure 2 illustrates approximately where they are located in the island.

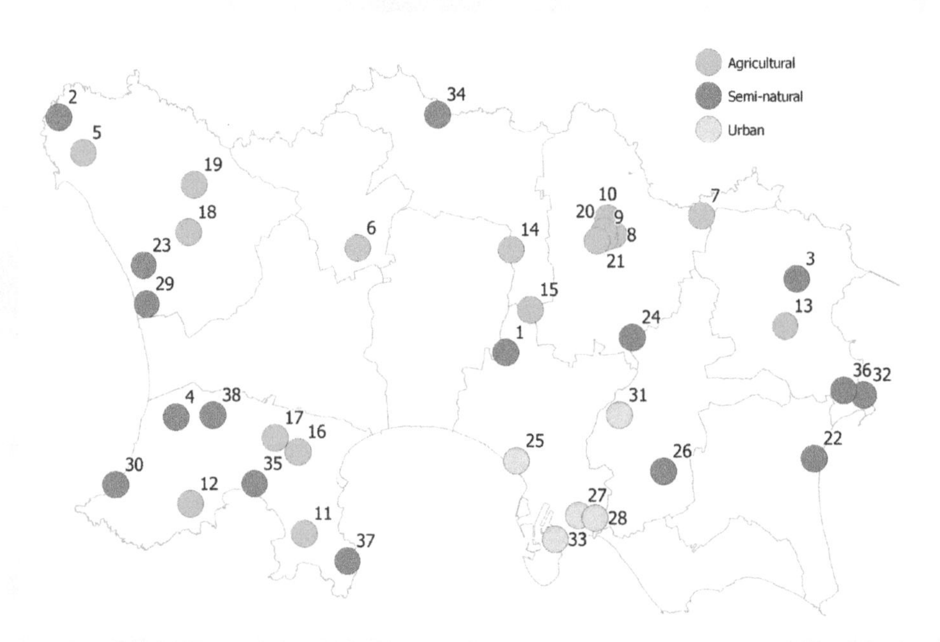

Figure 2. The location of the JBMS transects together with a classification based on their dominant habitat types (see Section 3.2). The location numbers refer to Table 2.

- Section Two -

Jersey's Butterfly Species: Individual Accounts

2.1 - Species Recorded by the JBMS

This section contains information on the 34 butterfly species recorded by JBMS volunteers between 2004 and 2013. Provided is a summary of the JBMS statistics for each species together with a discussion on its status within Jersey. Much of the pre-JBMS information comes from Long (2009) and Shaffer (2008). Although species' photographs are used, they are for illustrative rather than identification purposes (see Bibliography for a list of identification guides). The following information is provided:

Change
The percentage change in the butterfly species' population between 2004 and 2013. Figures are given for Jersey and the UK over the same time period (see also Section 3.1).

Total Counted
The total number of recorded reports of the species by JBMS volunteers between 2004 and 2013.

No. JBMS Transects
The total number of JBMS transects on which the species was recorded between 2004 and 2013.

Wider-countryside or Habitat Specialist
Indicates whether the species is considered to be a wider-countryside (or generalist) species or a habitat specialist (see Section 3.4 for more details).

Population Trend
The indices of abundance are generated using the software packages Transect Walker and TRIM (see Section 1.2). The 10 year UK and Jersey trends are for most of the JBMS species. (The UK data were kindly provided by Dr Tom Brereton, Butterfly Conservation). For some of the rarer species reported during the JBMS there are not enough records to run this analysis.

Butterfly and Larval Food Plant Distribution
A map showing the JBMS transect locations where the species was recorded (circles) together with their recorded larval food plants (stars). The species' abundance (circles) is the average number of individuals recorded per kilometre while the larval food plants (stars) are the number of records in a 1 kilometre square between 2004 and 2013. The size of the circles and stars are proportionate to the figures they represent.

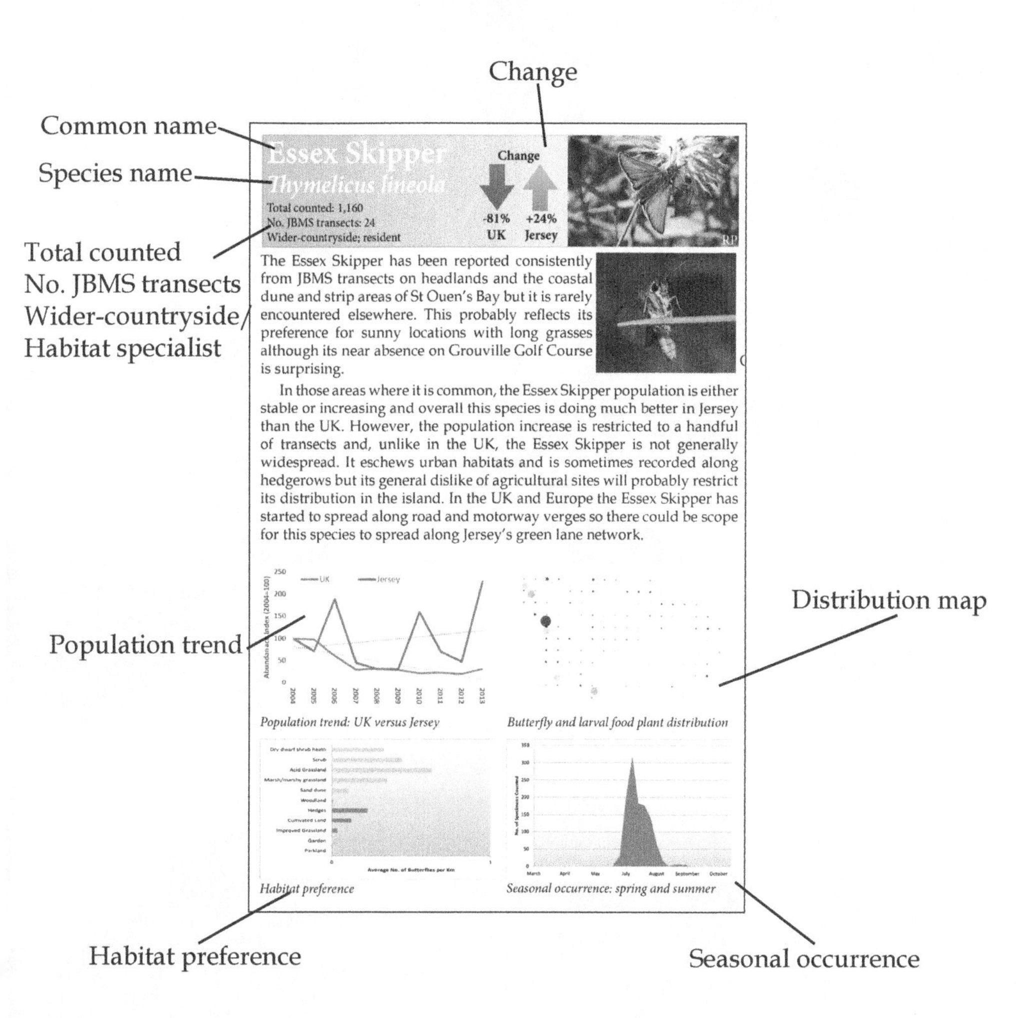

Essex Skipper
Thymelicus lineola
Total counted: 1,160
No. JBMS transects: 24
Wider-countryside; resident

Change
-81% UK
+24% Jersey

The Essex Skipper has been reported consistently from JBMS transects on headlands and the coastal dune and strip areas of St Ouen's Bay but it is rarely encountered elsewhere. This probably reflects its preference for sunny locations with long grasses although its near absence on Grouville Golf Course is surprising.

In those areas where it is common, the Essex Skipper population is either stable or increasing and overall this species is doing much better in Jersey than the UK. However, the population increase is restricted to a handful of transects and, unlike in the UK, the Essex Skipper is not generally widespread. It eschews urban habitats and is sometimes recorded along hedgerows but its general dislike of agricultural sites will probably restrict its distribution in the island. In the UK and Europe the Essex Skipper has started to spread along road and motorway verges so there could be scope for this species to spread along Jersey's green lane network.

Population trend: UK versus Jersey

Butterfly and larval food plant distribution

Habitat preference

Seasonal occurrence: spring and summer

Seasonal Occurrence

A weekly sum of counts made across the ten-year period plotted between April and September. Obvious peaks in the graph will generally come after the chrysalis stage in the butterfly's life cycle but may also represent influxes of migrant adults or the awakening of overwintering adults. The timing of the peaks in Jersey species generally coincides with those observed for butterflies in the UK. (NB The chart legend does not list all the month names on it but they are represented by the data.)

Habitat Preference

Because each transect section has an associated Phase 1 habitat classification, it is possible show a butterfly's abundance within each habitat type. This is expressed as the average number of butterflies sighted per kilometre walked. These graphs give an indication of the habitat preference for each species. The green columns are semi-natural habitats; the blue ones are agricultural habitats; and the orange ones are urban habitats.

Essex Skipper

Thymelicus lineola

Total counted: 1,160
No. JBMS transects: 24
Wider-countryside; resident

The Essex Skipper has been reported consistently from JBMS transects on cliff-tops and the coastal dune and strip areas of St Ouen's Bay but it is rarely encountered elsewhere. This probably reflects its preference for sunny locations with long grasses although its scarcity on Grouville Golf Course is surprising.

In those areas where it is common, the Essex Skipper population is either stable or increasing and overall this species is doing much better in Jersey than the UK. However, the population increase is restricted to a handful of transects and, unlike in the UK, the Essex Skipper is not generally widespread. It eschews urban habitats and is sometimes recorded along hedgerows but its general dislike of agricultural sites will probably restrict its distribution in the island. In the UK and Europe the Essex Skipper has started to spread along road and motorway verges so there could be scope for this species to spread along Jersey's green lane network.

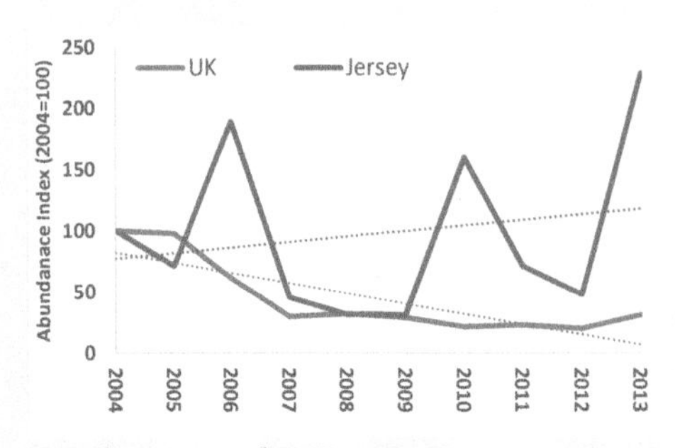

Population trend: UK versus Jersey

Butterfly and larval food plant distribution

Habitat preference : No. butterflies per Km

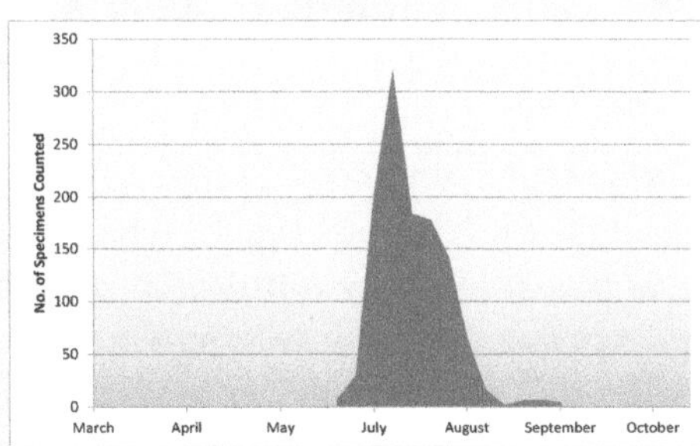

Seasonal occurrence: spring and summer

Large Skipper

Ochlodes sylvanus

Total counted: 1,041
No. JBMS transects: 31
Wider-countryside; resident

The Large Skipper may be found in low numbers across Jersey but its preferred habitat seems to be scrubby, grassy areas such as are found on cliff-tops and field verges. It also turns up in gardens and parks but is rarely seen on cultivated agricultural land.

There has been a steep increase in the Large Skipper population during the JBMS monitoring but this has probably been skewed upwards by particularly high counts made at Les Landes during 2013. With the Les Landes figures removed the population trend elsewhere is stable or possibly even slightly declining.

As a wider countryside butterfly associated with field margins and roadside verges, the Large Skipper should be suited to Jersey. The Large Skipper's life cycle is reliant on grass species (especially Cock's Foot, *Dactylis glomerata)* and initiatives to improve the management of verges in favour of insects may benefit species such as this.

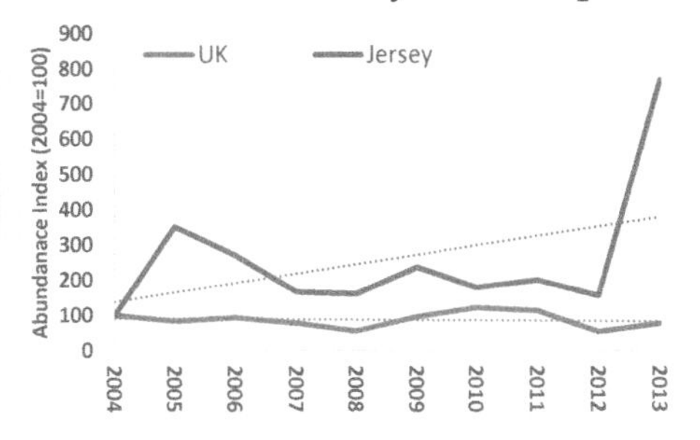

Population trend: UK versus Jersey

Butterfly and larval food plant distribution

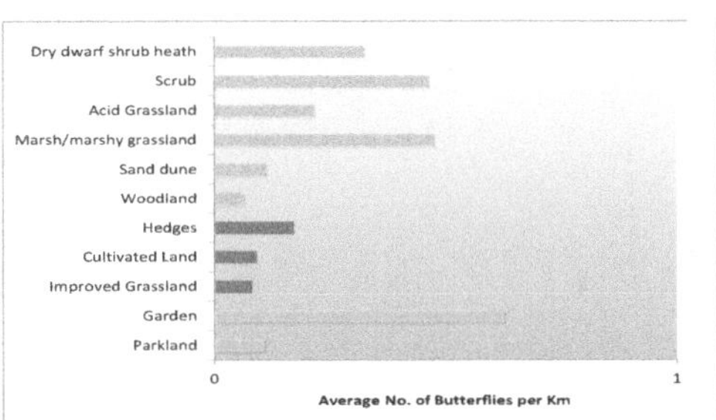

Habitat preference: No. butterflies per Km

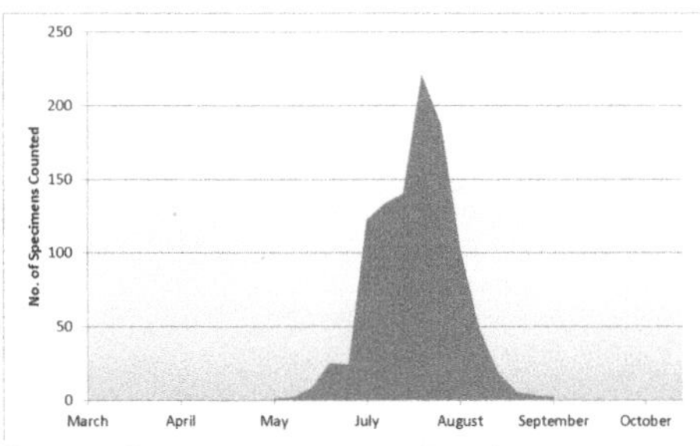

Seasonal occurrence: spring and summer

Swallowtail
Papilio machaon

Total counted: 84
No. JBMS transects: 9
Habitat specialist; resident(?)

Although recorded from nine different JBMS transects, most of the 84 records for this iconic butterfly come from just two locations: Grouville Golf Course and Victoria Tower, although its caterpillars have been found elsewhere. It is probable that there were resident colonies of Swallowtails at these two transects between 2006 and 2008 (Grouville) and from 2008 to 2011 (Victoria Tower). However, there has been no sighting at either location since 2011.

The Jersey Swallowtail is the continental variety *gorganus* which is more active than the British variety *brittannicus*. This makes a comparison of their population trend uncertain. The caterpillar is particular about its food plant and in Jersey it has only been observed feeding on Wild Carrot (*Daucus carota*) and Fennel (*Foeniculum vulgare*) . The decline of the Swallowtail in Jersey has been anecdotally ascribed to over enthusiastic site management leading to the destruction of its food plant. With management the species might re-establish itself but at present its status on the island is uncertain.

Population trend: UK versus Jersey

Butterfly and larval food plant distribution

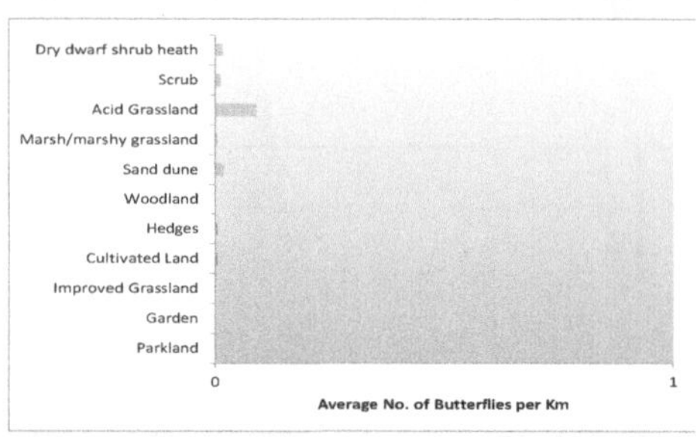

Habitat preference: No. butterflies per Km

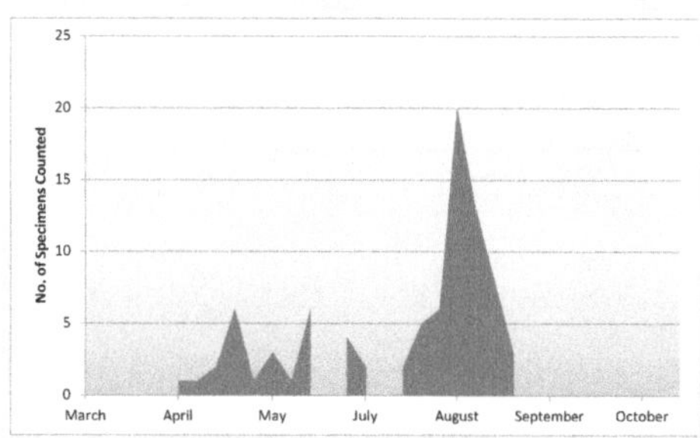

Seasonal occurrence: spring and summer

Clouded Yellow

Colias croceus

Total counted: 525
No. JBMS transects: 26
Migrant

Change

-88% **UK**

-91% **Jersey**

RP

The Clouded Yellow is a distinctive butterfly that migrates northwards from the Mediterranean and may be seen in Jersey during the late summer and autumn. Although reported every year, some summers are better than others with 2006, 2009 and 2013 being particularly good years.

The Clouded Yellow may be seen almost anywhere on Jersey but the most consistent reports are in open areas such as cliff-tops, grasslands and sand dunes. There is no evidence to suggest that this species is overwintering here although it is possible that some of the late summer reports are of locally bred individuals produced from early seasonal migrant parents.

The northerly range of the Clouded Yellow has increased markedly in recent decades suggesting that a warming climate may be pushing its distribution further north. As a migrant the Clouded Yellow's presence in Jersey is liable to be dependent on conditions on the continent.

Population trend: UK versus Jersey

Butterfly and larval food plant distribution

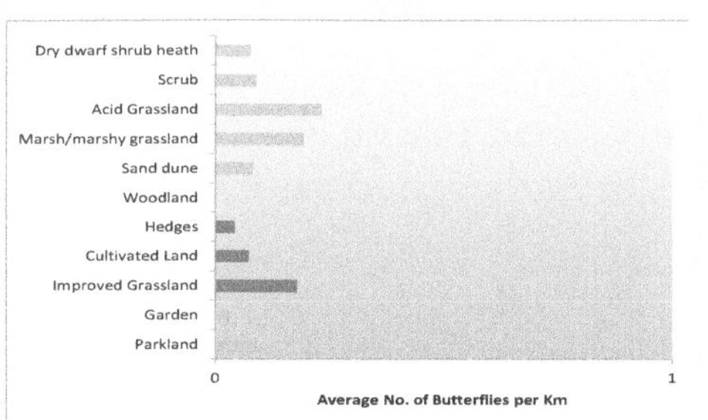

Habitat preference: No. butterflies per Km

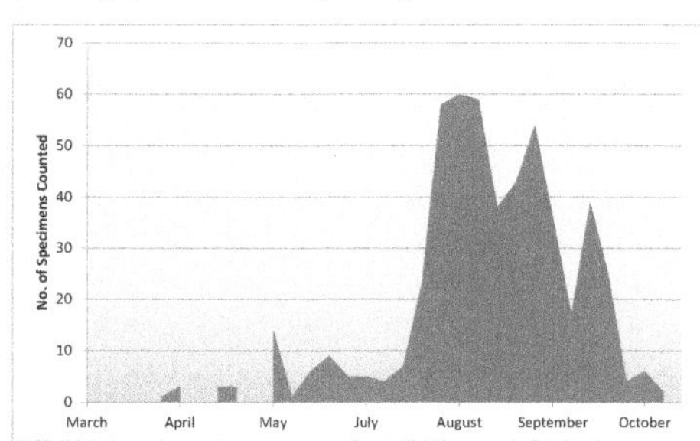

Seasonal occurrence: spring and summer

Pale Clouded Yellow

Colias hyale

Total counted: 5
No. JBMS transects: 2
Migrant

CCL

A southern European migrant species that is rarely reported from Jersey. The JBMS has one sighting from Fern Valley and four from Les Blanches Banques making it one of the rarest species recorded during the ten years of monitoring

During the summer of 1945 several individuals were found on the island but otherwise this beautiful migrant species is rather rare and random in its occurrence.

Roger Long and Ian Everson (pers. comm.) note that this is a very difficult species to identify by sight alone and that as such caution must be applied to any recent Jersey records. This includes the JBMS records which could be misidentifications of the pale *helice* form of the Clouded Yellow.

Orange-tip

Anthocharis cardamines

Total counted: 54
No. JBMS transects: 7
Wider countryside; resident

RP

A distinctive but uncommon species in Jersey which occurs in localised colonies. In the JBMS Orange-tips were only consistently reported from St Catherine's Woods with isolated reports coming from cliff-top and valley sites.

RP

In the UK the Orange-tip has expanded its range in recent decades with local populations remaining relatively stable. The same may be true in Jersey where this species was rare until 1980s after which local colonies were established. These colonies are restricted to a few locations and although in the UK and Europe the Orange-tip is commonly associated with green lanes, hedgerows and gardens, this does not seem to be the case in Jersey.

It is probable that the Orange-tip will remain localised in Jersey as its main food plant (Cuckooflower; *Cardamine pratensis*) has a restricted distribution. However, the Orange-tip is a species that, given the right circumstances, could potentially extend its distribution.

Brimstone

Gonepteryx rhamni

Total counted: 45
No. JBMS transects: 14
Wider countryside; vagrant(?)

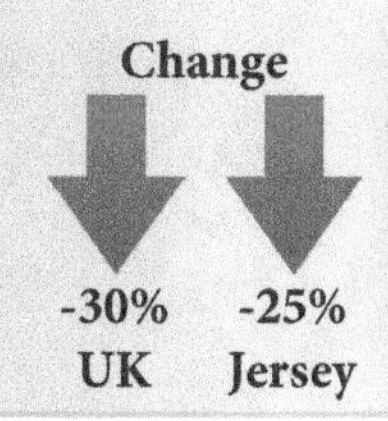

The Brimstone butterfly was rarely recorded during the JBMS monitoring with a total of 45 records made across 14 transects. In Jersey the Brimstone displays no habitat preference and reports come from semi-natural, agricultural and urban areas.

In the UK and Europe the Brimstone overwinters as an adult and is one of the first butterflies to emerge in spring. In Jersey the Brimstone's sporadic distribution and a lack of reports suggests that it is not resident but more probably an occasional vagrant from the continent.

The Brimstone is a frequenter of woodland edges and hedgerows and its distribution generally follows its larval food plants Alder Buckthorn (*Frangula alnus*) and Buckthorn (*Rhamnus cathartica*). These plants are absent from the island making it unlikely that this species will breed on the island although nineteenth-century records hint that the species may once have been more common and possibly even resident.

Population trend: UK versus Jersey

Butterfly and larval food plant distribution

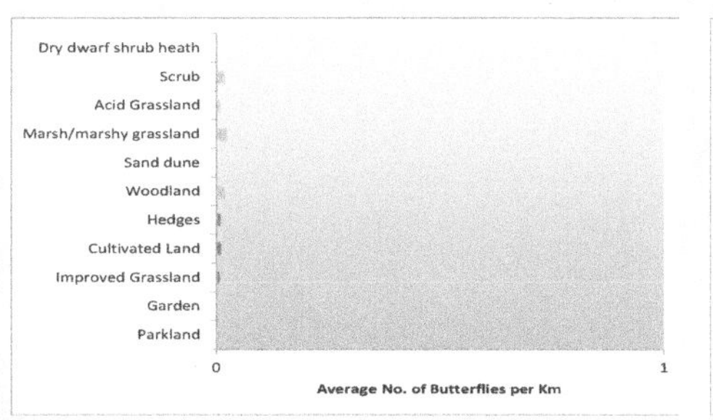

Habitat preference: No. butterflies per Km

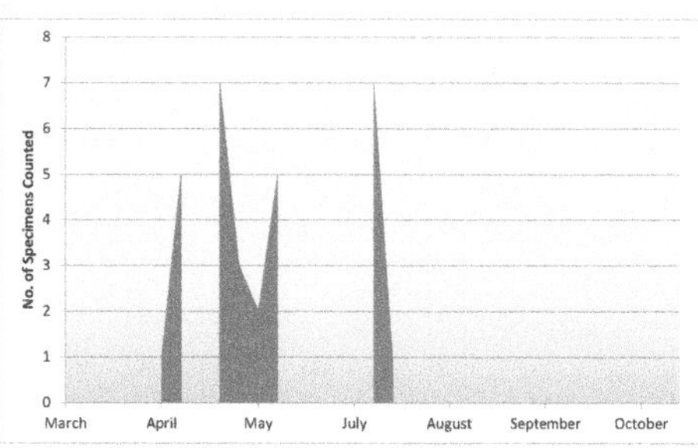

Seasonal occurrence: spring and summer

Large White

Pieris brassicae

Total counted: 10,073
No. JBMS transects: 38
Wider countryside; resident

Unloved by gardeners for its destructive caterpillars, the Large White is a ubiquitous butterfly that can be seen on every habitat type in Jersey. It is also one of the few butterflies that is commonly recorded in parks, gardens and agricultural land.

The Large White is an active flier that can cover long distances and hence it may wander almost anywhere including urban areas such as town centres. Its caterpillars feed on brassicas and pupation often occurs on man-made structures such as fences and walls. As such it is one of the few species to have benefited from the gradual industrialisation of European cities and farmland. Even so, the JBMS data suggests that the Large White has declined steeply in Jersey during the past decade whereas numbers in the UK are generally stable.

Population trend: UK versus Jersey

Butterfly and larval food plant distribution

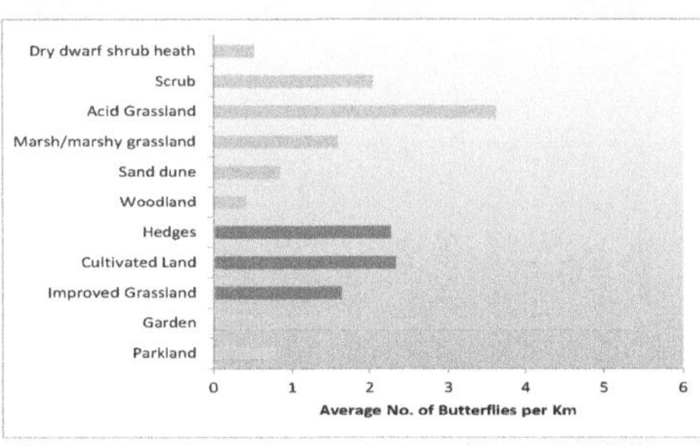

Habitat preference: No. butterflies per Km

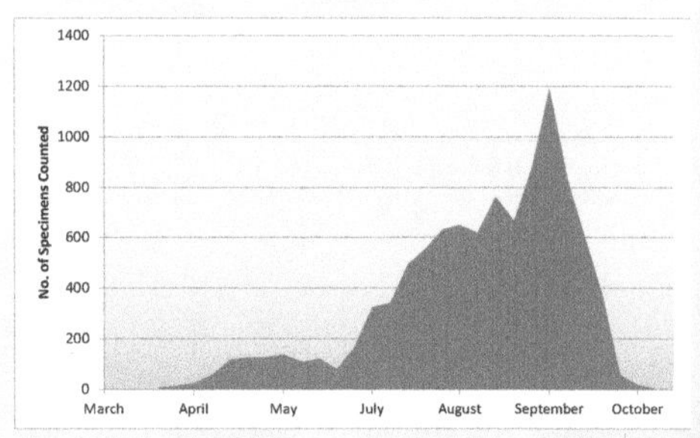

Seasonal occurrence: spring and summer

Small White

Pieris rapae

Total counted: 14,791
No. JBMS transects: 38
Wider countryside; resident

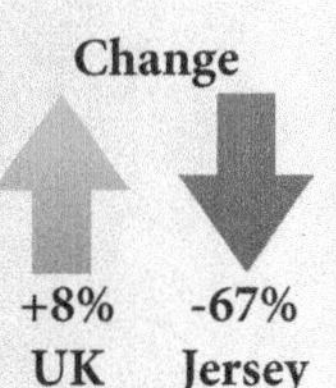

The Small White is a common and highly mobile species that has been recorded in reasonable numbers from all the JBMS transects. It is the only Jersey butterfly which has been more frequently recorded in agricultural areas than anywhere else.

Although the Small White is one of Jersey's commonest butterflies, the JBMS monitoring has recorded a steep decline in its population, especially since 2008. This stands in contrast to a moderate population increase seen across the UK.

As a species that is linked to agriculture, this may be a consequence of intensive farming practices (although UK Small Whites will be subject to this as well). Its principal larval food plant (brassicas) does not seem to have decreased in abundance during the JBMS monitoring period. Future monitoring data from the JBMS should help to determine if the Small White's decline is a matter of concern.

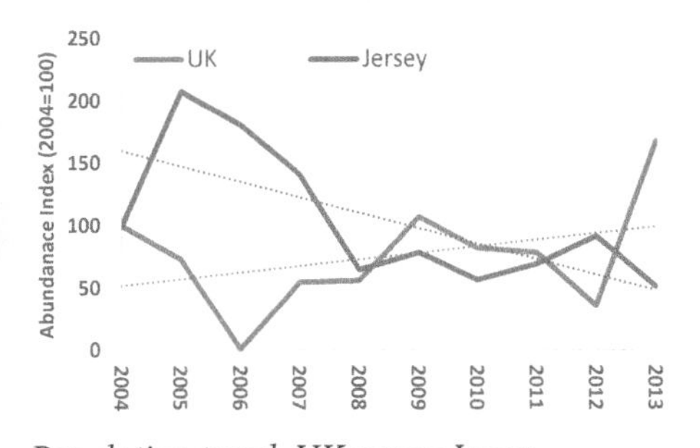

Population trend: UK versus Jersey

Butterfly and larval food plant distribution

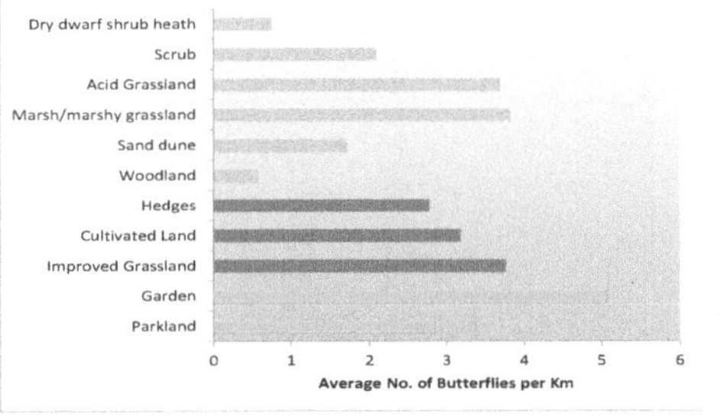

Habitat preference: No. butterflies per Km

No. of Specimens Counted
1600
1400
1200
1000
800
600
400
200
0
March April May July August September October

Seasonal occurrence: spring and summer

Green-veined White

Pieris napi

Total counted: 878
No. JBMS transects: 25
Wider countryside; resident

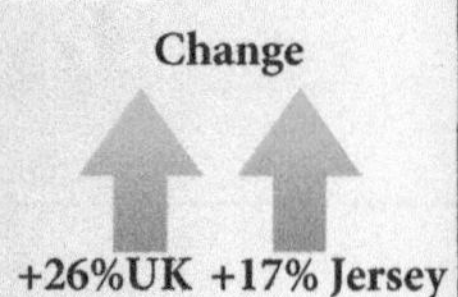

The Green-veined White has been reported from most of Jersey's habitat types but it was only consistently recorded on three JBMS transects. On two of these transects (Field C and West Park) regular reports have ceased while at Les Landes reports increased steeply in 2012.

The recent high numbers recorded at Les Landes mask a decline in the Green-veined White elsewhere in the island. This is in contrast to the UK, where the population is increasing.

As a butterfly that generally prefers slightly damper habitats, it is possible that the Green-veined White has colonies in areas that are under-represented by the JBMS, such as woodlands and wet meadows. However, based on the current analysis, this butterfly seems to be uncommon in Jersey with a restricted distribution which may make it vulnerable. Future JBMS monitoring may help to better define its status.

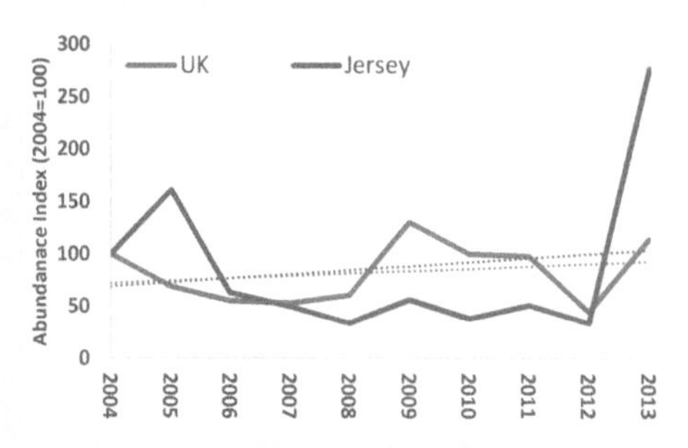

Population trend: UK versus Jersey

Butterfly and larval food plant distribution

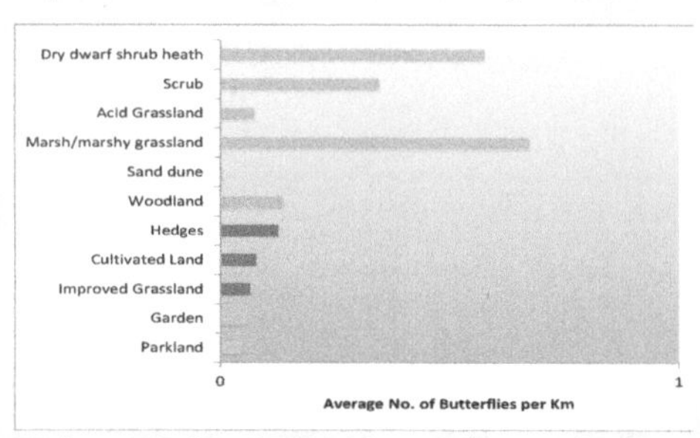

Habitat preference: No. butterflies per Km

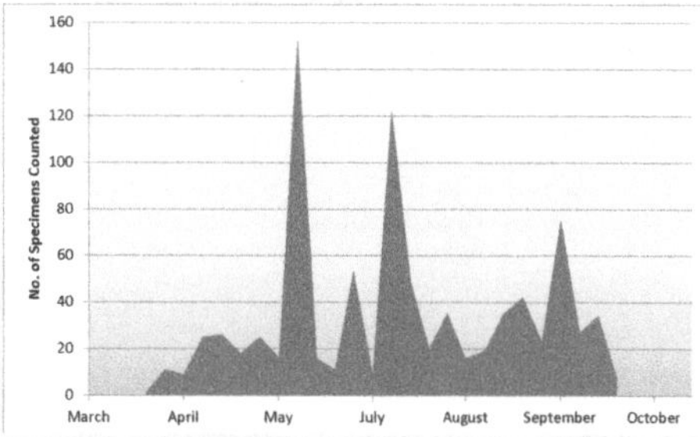

Seasonal occurrence: spring and summer

Green Hairstreak

Callophrys rubi

Total counted: 5,130
No. JBMS transects: 15
Habitat specialist; resident

The Green Hairstreak is a habitat specialist whose JBMS records are almost entirely restricted to the headland heath at Les Landes where it is common and its population is steeply increasing. The only other JBMS transects where it was regularly seen are at Les Blanches Banques and L'Oeillère Headland but in lower numbers.

Historically the Green Hairstreak seems to have been more widespread on Jersey's cliffs and escarpments and, although it is doing well at Les Landes, this may represent its last refuge on the island which suggests that its population may be vulnerable.

Even along the transect at Les Landes the Green Hairstreak distribution is not uniform with the species being particularly associated with areas of dry dwarf shrub heath. Further investigations into the Green Hairstreak's occurrence and ecology in Jersey are desirable to see if any conservation measures are needed to safeguard its future.

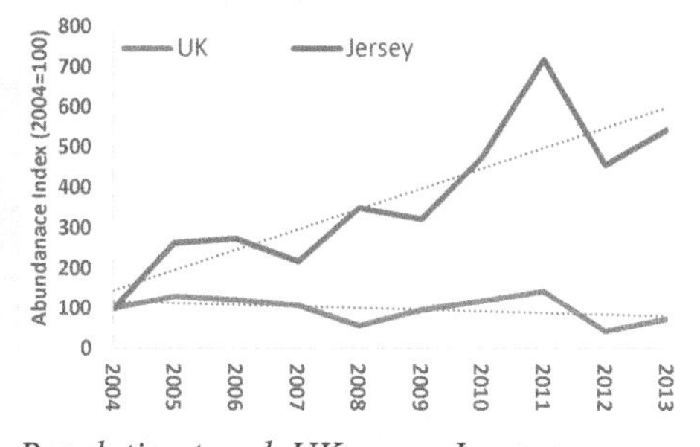

Population trend: UK versus Jersey

Butterfly and larval food plant distribution

Dry dwarf shrub heath
Scrub
Acid Grassland
Marsh/marshy grassland
Sand dune
Woodland
Hedges
Cultivated Land
Improved Grassland
Garden
Parkland
0 1 2 3 4 5 6 7 8 9 10
Average No. of Butterflies per Km

Habitat preference: No. butterflies per Km

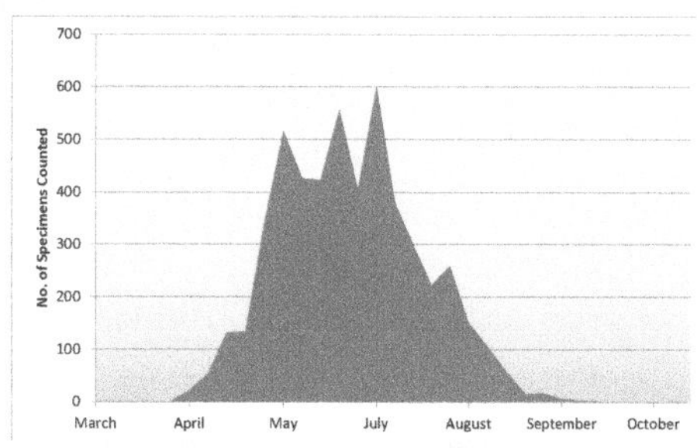

Seasonal occurrence: spring and summer

Purple Hairstreak

Favonius quercus

Total counted: 24
No. JBMS transects: 3
Wider countryside; resident

The Purple Hairstreak is one of the rarer JBMS butterfly species but it was recorded consistently (if infrequently) at the transects in West Park, Les Landes and St Catherine's Woods.

The Purple Hairstreak is a high flying butterfly associated primarily with tree canopies (especially oak trees). The JBMS monitoring technique, which records butterflies flying close to the volunteer, does not suit high flying species and so the Purple Hairstreak may be under-represented.

Other island records suggest that the Purple Hairstreak is locally common but was possibly more widespread historically than at present. The recognition of its under-representation in UK butterfly monitoring has led to deliberate searches for this species. Results from this suggest that the Purple Hairstreak is more widespread and abundant than had been previously supposed.

Jersey's Purple Hairstreak population is probably stable but this species' dependence on oak trees makes it potentially vulnerable to changes in hedgerow and woodland management.

White-letter Hairstreak

Satyrium w-album

Total counted: 42
No. JBMS transects: 5
Wider countryside; resident

The White-letter Hairstreak is one of Jersey's rarer butterflies and was recorded, mostly individually, from just five JBMS transects. In 2012 it was regularly recorded at Les Landes but then in 2013 was seen just twice.

This pattern seems to be reflected in its historical recording in Jersey where several reports from a single location will be followed by years where it is rarely seen at all.

The White-letter Hairstreak seems to have been more widespread between the 1960s and 1980s until the arrival of Dutch Elm Disease all but wiped out the exclusive food plant of its caterpillar. The gradual reintroduction of Elm trees to the island since the 1990s may help to encourage its re-establishment and spread.

Long-tailed Blue

Lampides boeticus

Total counted: 17
No. JBMS transects: 1
Migrant

RP

The Long-tailed Blue is a migrant species from southern Europe that has been reported occasionally from Jersey and which, according to Long (2009) has become more commonly seen on the island during the past thirty years.

All the JBMS records are from the transect at West Park where a small but consistent number of Long-tailed Blues were reported in 2006 followed by individual reports in 2009, 2011 and 2013. These sighting came at the end of the summer which suggests they are of individuals that migrated in from Europe but it is probable that the 2006 records were of a temporary breeding population (Ian Everson, pers. comm.). In 2015 Long-tail Blues were reported from a number of locations around Jersey but it is probably incapable of surviving an average winter here and is unlikely to be permanently resident.

Ringlet

Aphantopus hyperantus

Total counted: 3
No. JBMS transects: 2
Wider countryside; vagrant(?)

CCL

Although common in the UK, the Ringlet is a rare butterfly in Jersey probably because of a lack of suitable habitat. The JBMS has three ringlet records, one from L'Ouaisné and two from Field K; other Jersey records are sporadic and isolated and there is no indication that this species is resident here.

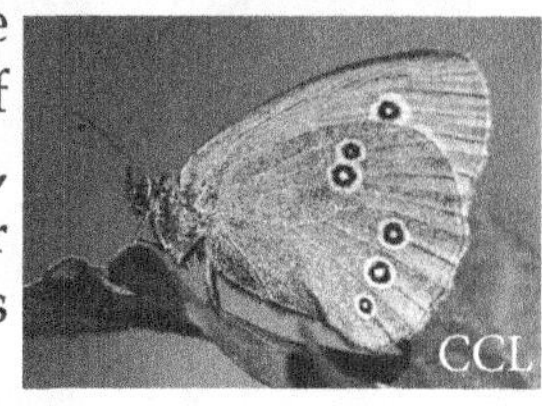
CCL

The Ringlet likes damp and shady vegetated areas and does not do well in open, drier conditions. Other than within some wooded valleys, its preferred habitats are largely absent from Jersey which, combined with a dislike of intensive agriculture, makes the island a generally unfavourable place for this species. Occasional reports are to be expected but the Ringlet is unlikely to become a widely established species in Jersey.

There is the possibility that the JBMS Ringlet reports may represent misidentifications of the male Meadow Brown butterfly (Ian Everson, pers. comm.). Specimens have been taken in the past (e.g. 1946 and 1951) so the Ringlet is not unknown on the island but photographic confirmation of the species' current presence on the island is desirable.

Small Copper

Lycaena phlaeas

Total counted: 1,960
No. JBMS transects: 34
Wider countryside; resident

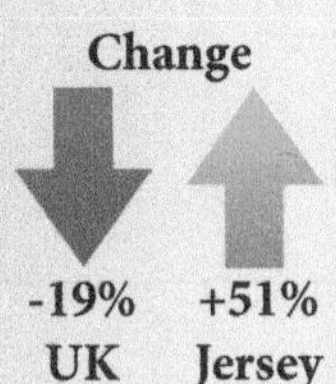

The Small Copper was most commonly reported on open semi-natural sites but it was also regularly recorded in agricultural areas, parks and woodlands. Jersey's temperate climate probably suits the Small Copper and it does not seem to suffer from the cold-inspired population collapses that have periodically occurred in some parts of the UK.

Jersey's Small Copper population is steeply increasing (in contrast to the UK) but as with several other local butterflies, this trend is restricted to a few semi-natural sites where it is common. Outside these areas reports are isolated and random and often associated with parks and hedgerows.

Although not threatened with imminent extinction, the Small Copper is possibly relying on localised breeding colonies in Jersey which could make the species vulnerable to changes in local habitat. In the UK the Small Copper is believed to have declined as a consequence of intensive farming and so it is therefore possible that this may have occurred in Jersey too.

Population trend: UK versus Jersey

Butterfly and larval food plant distribution

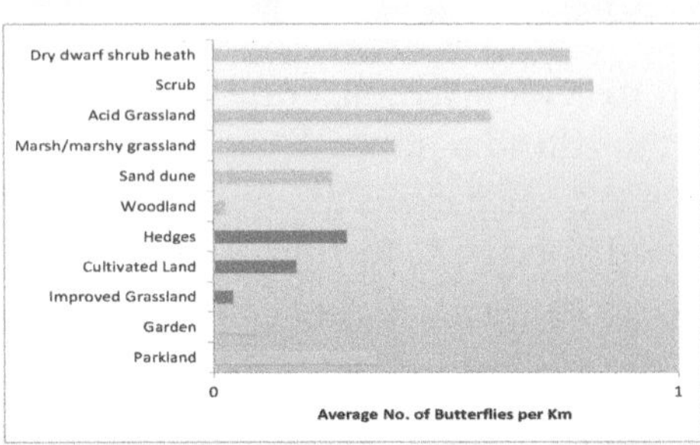

Habitat preference: No. butterflies per Km

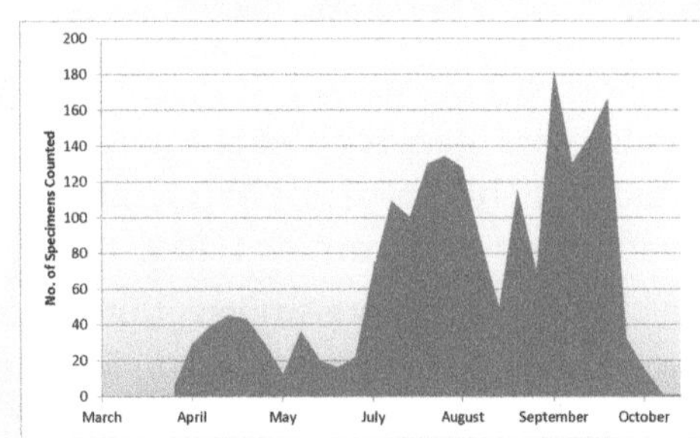

Seasonal occurrence: spring and summer

Brown Argus

Aricia agestis

Total counted: 234
No. JBMS transects: 20
Wider countryside; resident

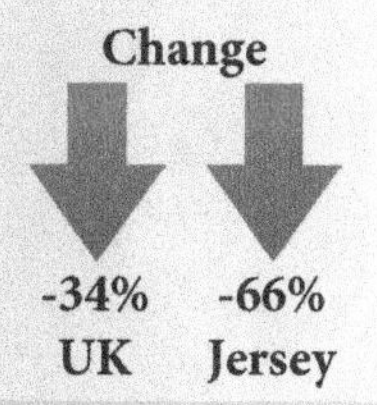

The Brown Argus was recorded across 20 JBMS transects but many of these were reports of isolated individuals. The only consistent reports are from Les Landes, Les Blanches Banques and Grouville Golf Course but even at these locations numbers are quite low. This is in contrast to historical reports which suggest that the species was more widespread and common until around the 1950s.

Jersey's Brown Argus population is steeply declining and is currently restricted to a few coastal heaths, grasslands and sand dunes. This distribution is possibly linked to that of its food plants which are more common on the semi-natural sites in the west of the island than elsewhere.

In the UK the Brown Argus has adapted to live in overgrown areas such as railway cuttings and abandoned land. This does not seem to be happening in Jersey and its long-term future on the island is far from assured. Further investigation is needed to see if conservation measures are needed to help assure the Brown Argus's future in Jersey.

Population trend: UK versus Jersey

Butterfly and larval food plant distribution

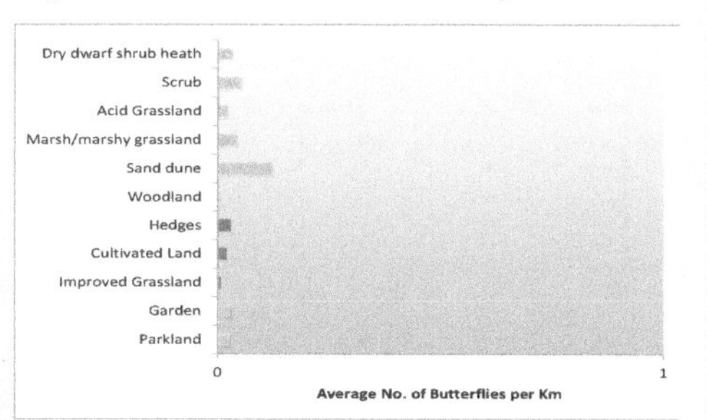

Habitat preference: No. butterflies per Km

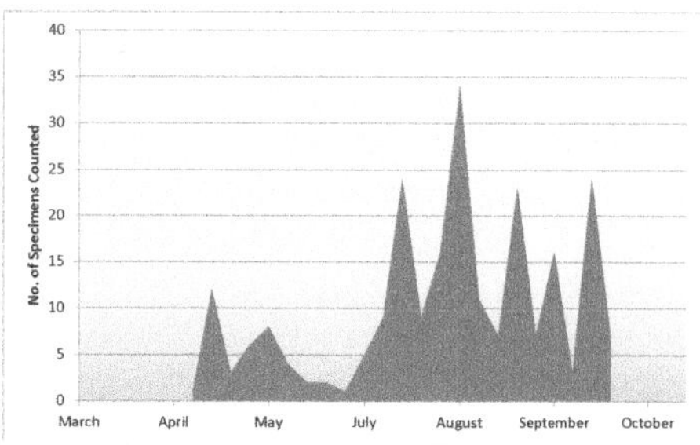

Seasonal occurrence: spring and summer

Common Blue

Polyommatus icarus

Total counted: 12,391
No. JBMS transects: 38
Wider countryside; resident

A common butterfly that may be seen almost anywhere in Jersey. Reports were made from most of Jersey's habitat types but a majority of the Common Blue records come from a handful of cliff-top transects at Les Landes, Field H or coastal areas at St Ouen and Grouville.

Although a wider countryside species, in Jersey the Common Blue does much better on semi-natural habitats than elsewhere. The JBMS monitoring suggests that while overall numbers are increasing, this species has good years and bad years and that on some transects, such as Grouville Golf Course, the population may be declining.

The Common Blue is sensitive to habitat change and in the UK it is considered to be a good indicator of the general state of biodiversity in the wider countryside. Ian Everson (pers. comm.) notes that the species' usual triple-brood is not reflected in the seasonal distribution graph (below); this might be due to the recent cold springs affecting its breeding cycle.

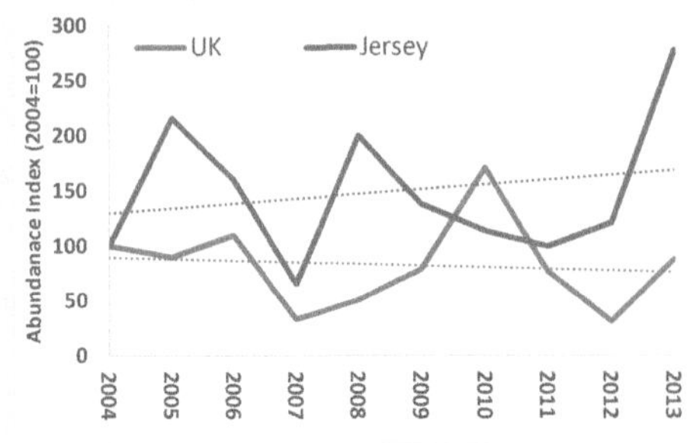

Population trend: UK versus Jersey

Butterfly and larval food plant distribution

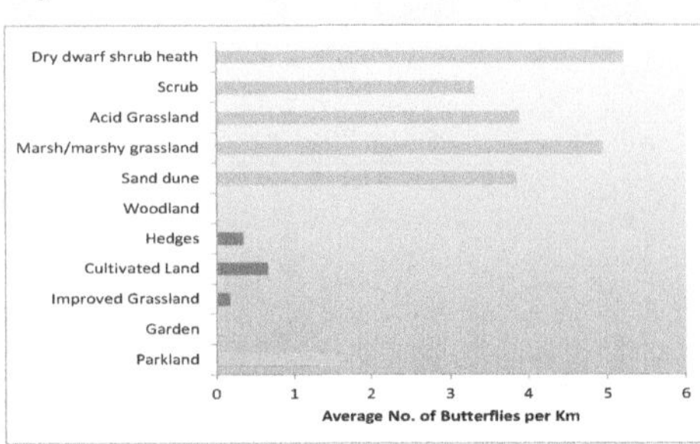

Habitat preference: No. butterflies per Km

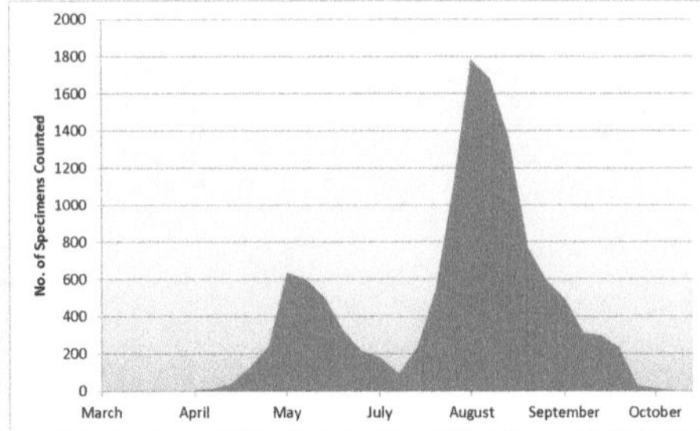

Seasonal occurrence: spring and summer

Holly Blue

Celastrina argiolus

Total counted: 1,290
No. JBMS transects: 32
Wider countryside

Jersey's Holly Blue population is increasing but, as with several other of the island's butterflies, its reports are concentrated into a small number of transects.

The Holly Blue was most frequently sighted on parklands (South Hill, West Park), golf courses (Grouville) and woodlands (St Catherine's Woods). It has also been reported from gardens, cliff-tops and sand dune areas but was more rarely seen in agricultural areas. The spring and summer peaks in its reports reflect its double brood.

Although not as vulnerable as some of Jersey's butterflies, populations are known to be intolerant of changes in climate and may also be parasitised by the wasp *Listrodomus nycthemerus*. As a frequenter of urban habitats, it is a species that could perhaps be encouraged by managing some areas of parkland and gardens in a butterfly-friendly manner.

Population trend: UK versus Jersey

Butterfly and larval food plant distribution

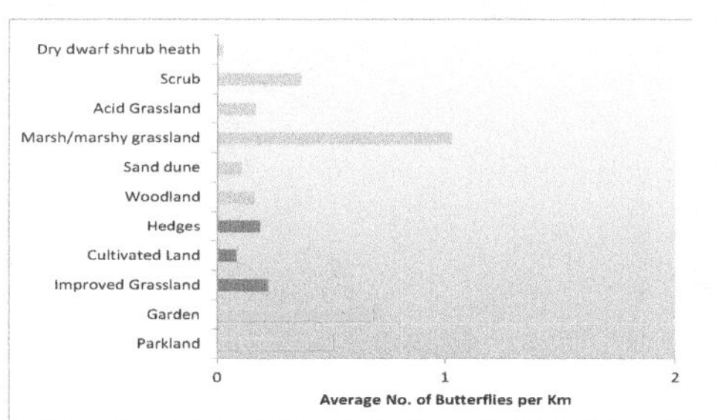

Habitat preference: No. butterflies per Km

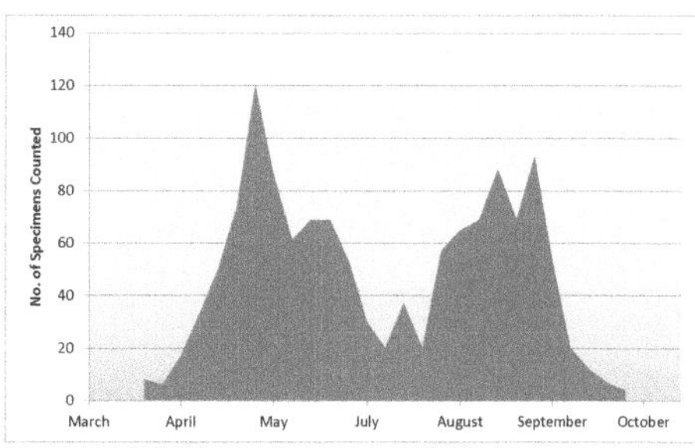

Seasonal occurrence: spring and summer

Red Admiral

Vanessa atalanta

Total counted: 3,755
No. JBMS transects: 38
Migrant/Wider countryside

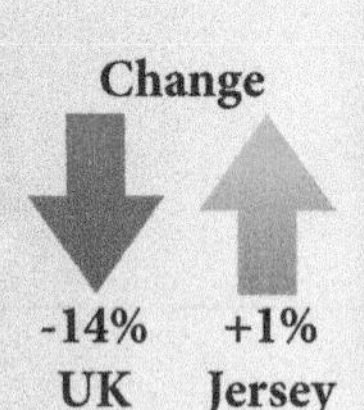

The Red Admiral is a distinctively large and colourful butterfly that is familiar to many gardeners. Although primarily a migrant from Europe, consistent reports from the late winter and early spring suggest that some adults are overwintering in the island. It is a strong-flyer which has been reported from every JBMS transect and it can be seen on almost every type of habitat including those in agricultural and urban areas.

The Red Admiral population in Jersey is stable and the species is not thought to be threatened. As a strong flier it may be seen almost anywhere in the island but is notable as a common visitor to parks and gardens. There is perhaps scope to assist this species by encouraging the planting of garden nectar plants and conservation management techniques in Jersey's parks.

Population trend: UK versus Jersey

Butterfly and larval food plant distribution

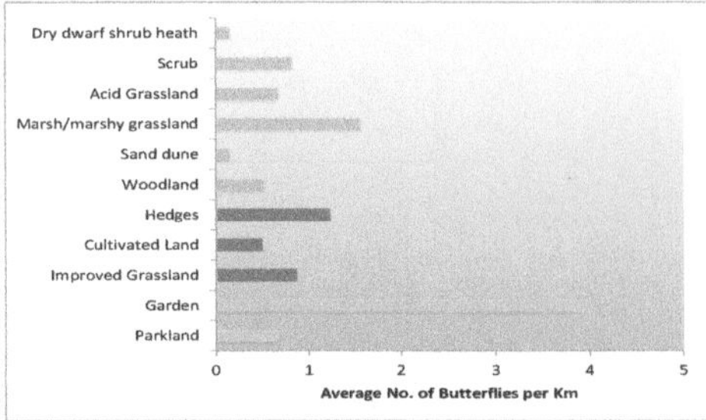

Habitat preference: No. butterflies per Km

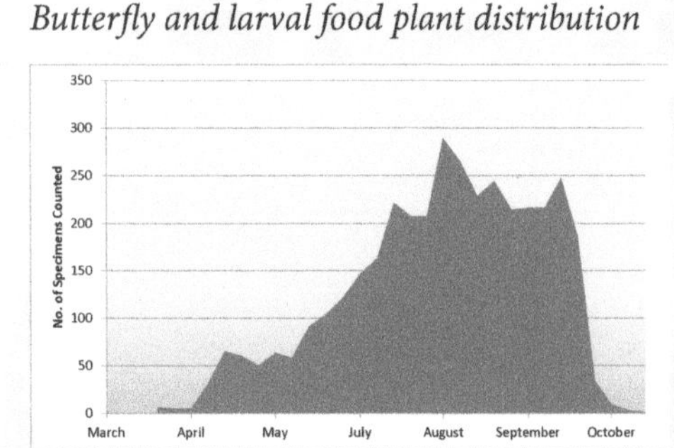

Seasonal occurrence: spring and summer

Painted Lady

Vanessa cardui

Total counted: 5,277
No. JBMS transects: 38
Migrant

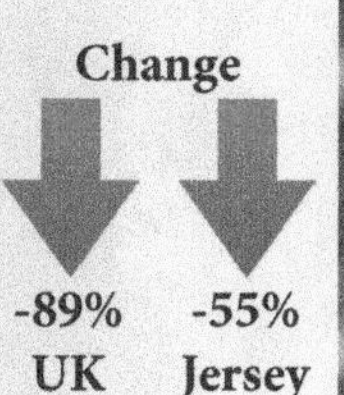

A large and colourful migrant from southern Europe whose populations can vary greatly from one year to the next with some years seeing the arrival of many thousands of adults. The year 2009 saw a spectacular European influx which received wide comment in the local and national media. This variability makes it hard to assess the population trend but in normal years the JBMS will receive between 50 and 100 reports.

The Painted Lady may be encountered anywhere but it is most commonly reported from gardens and open spaces including intensively farmed agricultural areas. Early season migrants can produce summer or autumn broods in Jersey but the adults are unable to survive over the winter. Given this, the state of Jersey's Painted Lady population is largely dependent on suitable breeding conditions in North Africa and southern Europe.

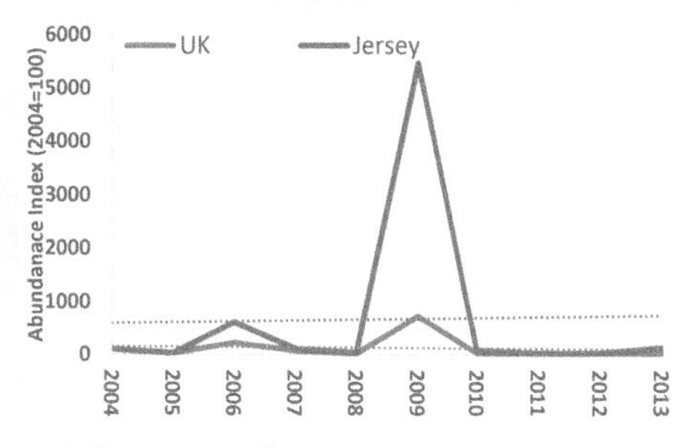

Population trend: UK versus Jersey

Butterfly and larval food plant distribution

Dry dwarf shrub heath
Scrub
Acid Grassland
Marsh/marshy grassland
Sand dune
Woodland
Hedges
Cultivated Land
Improved Grassland
Garden
Parkland
0
1
2
3
Average No. of Butterflies per Km

Habitat preference: No. butterflies per Km

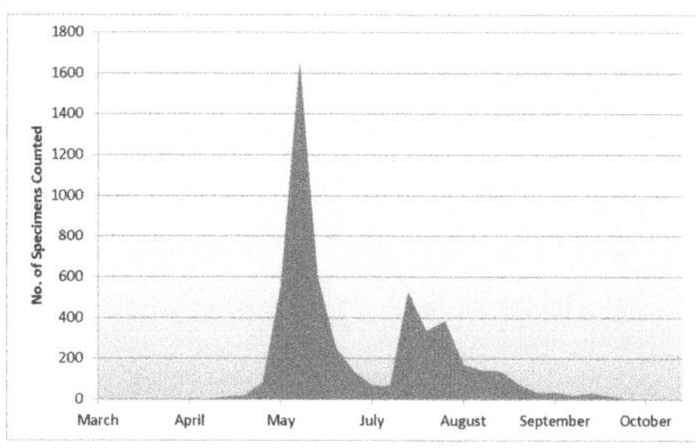

Seasonal occurrence: spring and summer

White Admiral

Limenitis camilla

Total counted: 101
No. JBMS transects: 8
Habitat specialist

The White Admiral is a very rare butterfly on Jersey. It was unrecorded until 1944 and only sporadically since. A small colony was reported from St Catherine's Woods in 2000 and there were consistent reports from there in 2006, 2008 to 2011 and 2013, suggesting that a colony was resident. However, numbers have declined since 2009 and reports from other transects are rare and isolated.

As a woodland butterfly whose eggs and caterpillar require Honeysuckle (*Lonicera* spp.), the White Admiral is unlikely ever to be widespread in Jersey and the JBMS data suggest that the colony in St Catherine's Woods may be difficult to sustain itself in the longer term. However, in the UK the White Admiral population is increasing and so there might be scope for its expansion into new woodland areas in Jersey, if suitable conditions (but especially its food plant) exist.

Large Tortoiseshell

Nymphalis polychloros

Total counted: 1
No. JBMS transects: 1
Habitat specialist

Regarded as functionally extinct in the UK, there was a small breeding population of Large Tortoiseshell butterflies in Jersey until just after World War Two.

The Large Tortoiseshell is still occasionally reported in Jersey and during the JBMS there was a confirmed report from St Catherine's Woods. Such isolated sightings are probably of migrant specimens or perhaps even of captive bred specimens that have been released into the wild.

As a butterfly that is associated with elm trees, it is probable that the effects of Dutch Elm Disease have restricted the chances of the Large Tortoiseshell becoming re-established in Jersey.

Small Tortoiseshell

Aglais urticae

Total counted: 230
No. JBMS transects: 23
Wider countryside

Change

-28%	+95%
UK	**Jersey**

The Small Tortoiseshell was a common species in Jersey until around 1999 when the population collapsed to near extinction (Long, 2009). There followed a slow recovery and within the JBMS the Small Tortoiseshell was recorded in relatively low numbers until 2013 when there was an upsurge in numbers, especially on semi-natural cliff-top and coastal transects. This late surge is reflected in the 95% population increase recorded by the JBMS.

As a generalist species that can pupate on man-made structures and which can live in a variety of habitats, including gardens and cultivated fields, the Small Tortoiseshell should probably be more common and widespread in Jersey than is currently the case. There is anecdotal evidence to suggest that it is still a regular visitor to gardens planted with buddleia and other butterfly-friendly plants but it does not seem to be thriving in the wider countryside. Further investigation is needed.

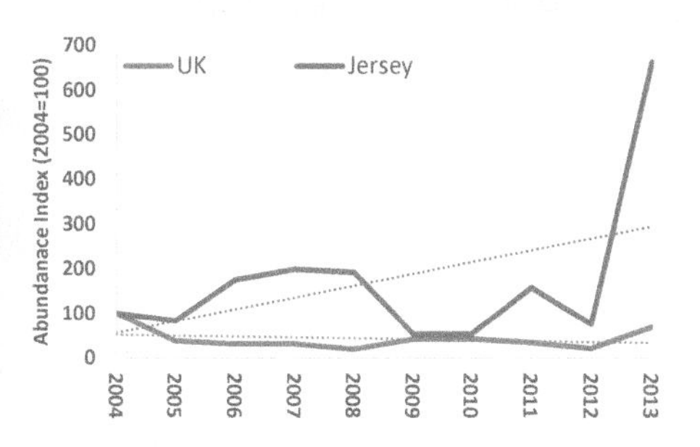

Population trend: UK versus Jersey

Butterfly and larval food plant distribution

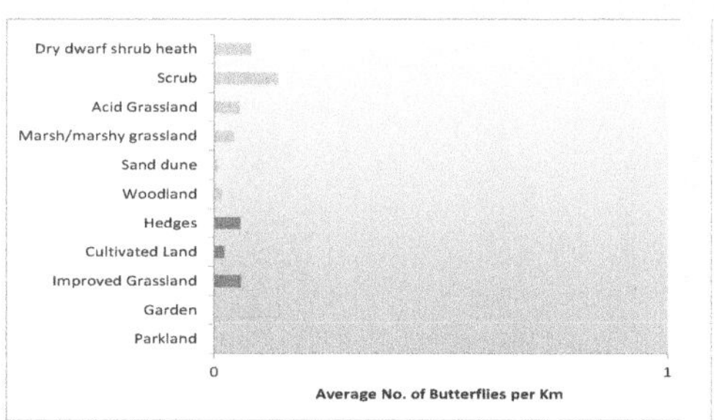

Habitat preference: No. butterflies per Km

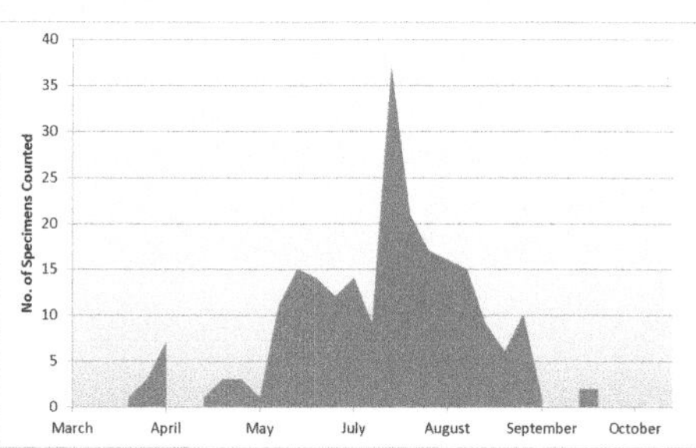

Seasonal occurrence: spring and summer

Peacock

Aglais io

Total counted: 788
No. JBMS transects: 33
Wider countryside; resident

An iconic butterfly whose overwintering adults are among the first butterflies to be reported by JBMS volunteers each year. Although not the commonest butterfly in Jersey, the Peacock is widespread and its size, distinctive patterning and fondness for garden plants makes it highly visible.

Although it may be encountered practically anywhere in Jersey, the Peacock shows a preference for open sunny sites such as cliff-tops, grasslands and gardens.

In contrast to the UK, the Peacock population in Jersey appears to have increased during the first few years of the JBMS and afterwards remained stable. The Peacock is not thought to be threatened but the adoption of butterfly friendly management areas in gardens and parks could help to encourage its numbers.

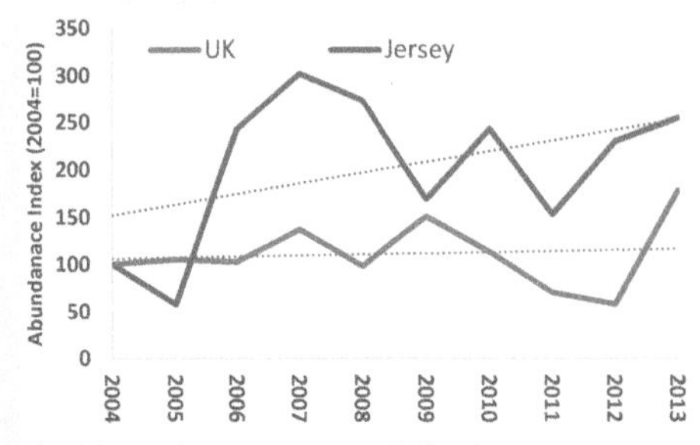

Population trend: UK versus Jersey

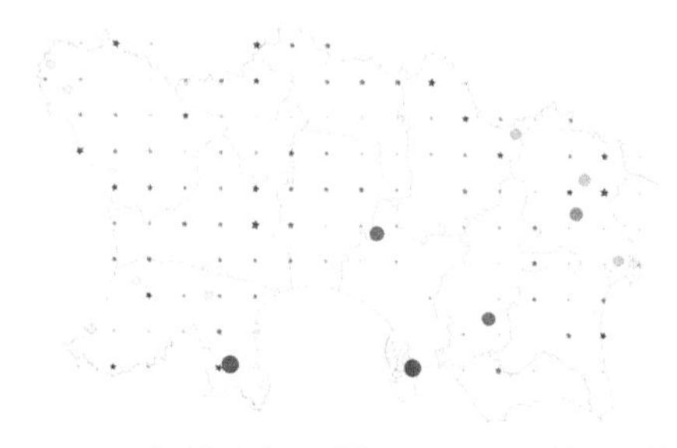

Butterfly and larval food plant distribution

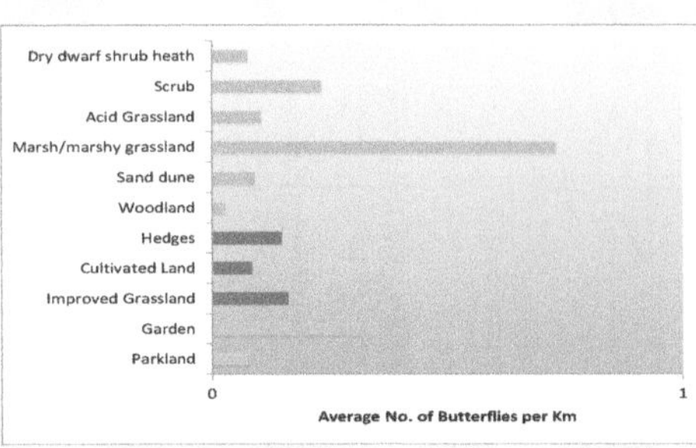

Habitat preference: No. butterflies per Km

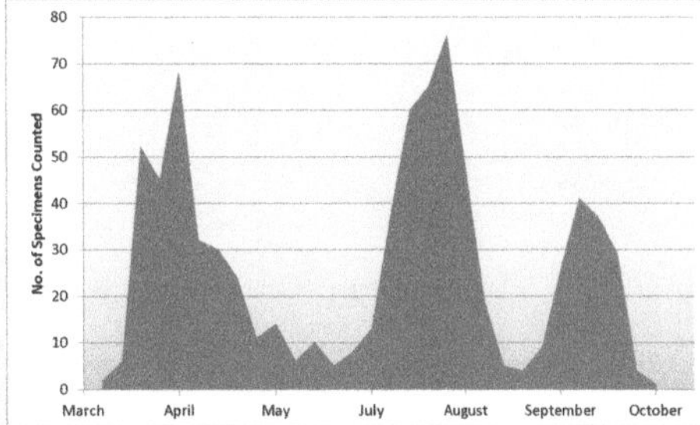

Seasonal occurrence: spring and summer

Comma

Polygonia c-album

Total counted: 495
No. JBMS transects: 27
Wider countryside; resident

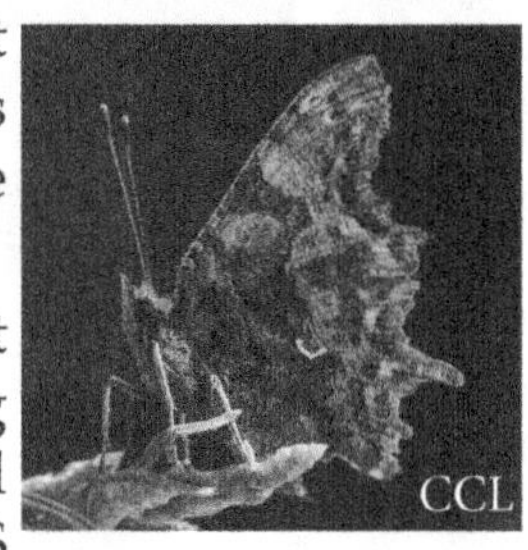

The Comma may be seen anywhere in Jersey but a majority of reports are restricted to St Catherine's Woods and Swiss Valley which reflects its preference for valley sites and open woodlands.

The Comma adult hibernates and is one of the first butterflies to be seen in the spring. It forms breeding populations in Jersey but these are probably restricted to a small number of sites and, while the JBMS suggests that the local population are stable, there is no sign of breeding populations emerging on new sites. This make this species vulnerable to change or the destruction of their preferred habitat.

In northern Europe the range of the Comma has expanded rapidly since the 1970s. This is probably due to warming climates allowing breeding colonies to establish themselves further north. The habitats that the Comma frequents are probably under-represented in the JBMS and further information on its distribution is desirable.

Population trend: UK versus Jersey

Butterfly and larval food plant distribution

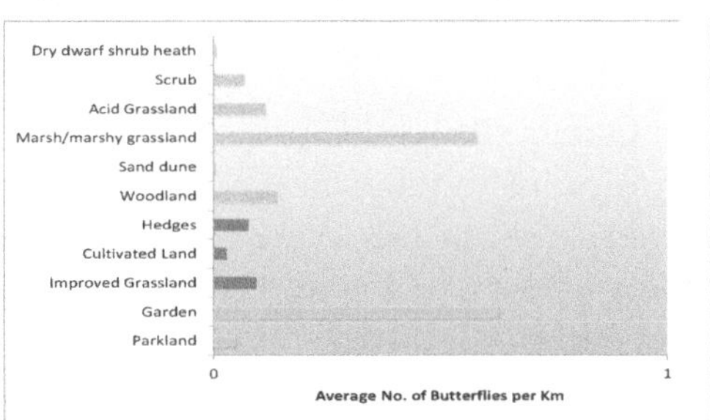

Habitat preference: No. butterflies per Km

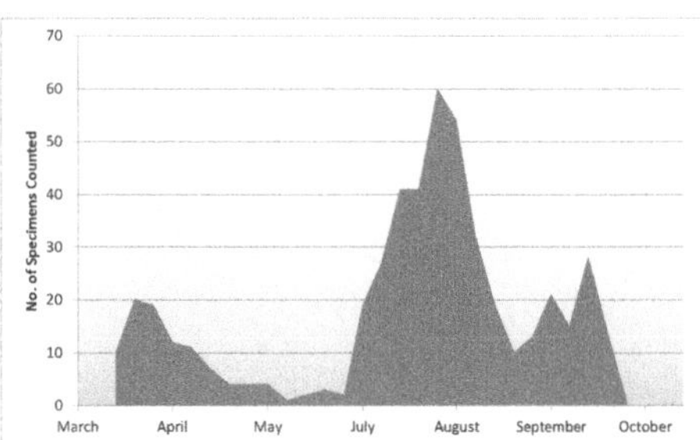

Seasonal occurrence: spring and summer

European Map

Araschnia levana

Total counted: 1
No. JBMS transects: 1
Migrant

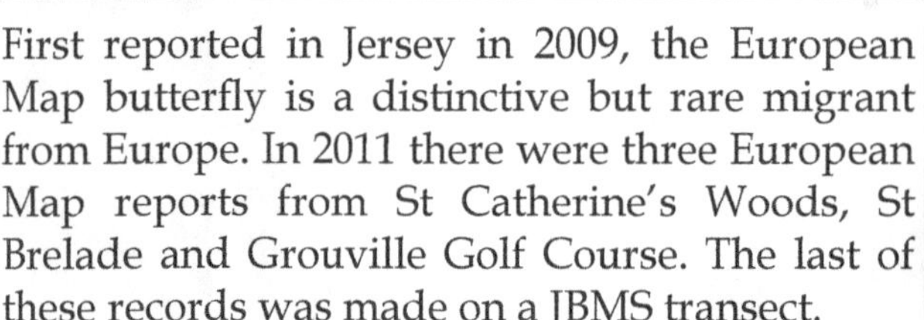

First reported in Jersey in 2009, the European Map butterfly is a distinctive but rare migrant from Europe. In 2011 there were three European Map reports from St Catherine's Woods, St Brelade and Grouville Golf Course. The last of these records was made on a JBMS transect.

The sighting of the first (top image) and second generation (right image) forms of this butterfly led to speculation that this butterfly had established a breeding population on the island but a lack of subsequent reports suggests that this has not been sustained. Jersey's experienced butterfly watchers are keeping an eye out for the European Map in the hope that it might become a permanent resident on the island.

Queen of Spain Fritillary

Issoria lathonia

Total counted: 6
No. JBMS transects: 3
Migrant/Wider countryside

Rarely recorded from the UK, the Queen of Spain Fritillary was first reported in Jersey in 1951 but until recently was only sporadically encountered.

Since 2004 the JBMS has recorded three reports from Les Blanches Banques and two from the urban sites in South Hill Park and one from West Park. This, plus other reports from outside of the JBMS, suggests that the Queen of Spain Fritillary is becoming a regular migrant to the island and it is even suspected to have been sporadically resident on Les Blanches Banques. However, its primary larval food plant (the Field Pansy; *Viola arvensis*) is not common or widespread in Jersey although it can sometimes utilise other species of *Viola*. The Queen of Spain Fritillary may sometimes visit gardens and increased awareness of the species may lead to more reports from the public.

Speckled Wood

Pararge aegeria

Total counted: 13,352
No. JBMS transects: 38
Wider countryside; resident

Change

-9%	-8%
UK	Jersey

The Speckled Wood is a very common and widespread butterfly in Jersey with reports coming from all JBMS transects. It can be seen on most habitats in Jersey where there is a reasonable amount of vegetation. The highest counts are in semi-natural areas and in some managed urban transects such as West Park and Green Street Cemetery.

As a wider countryside species that can overwinter as either a caterpillar or a chrysalis, the Speckled Wood is well-suited to Jersey's habitats and the population, while variable, seems to be stable. It is one of the few butterflies that is more common in the centre and east of the island than the west.

Jersey specimens are typical of those found in the southern UK and have wing spots that are orange in colour compared with the white forms seen in northern parts of Europe.

Population trend: UK versus Jersey

Butterfly and larval food plant distribution

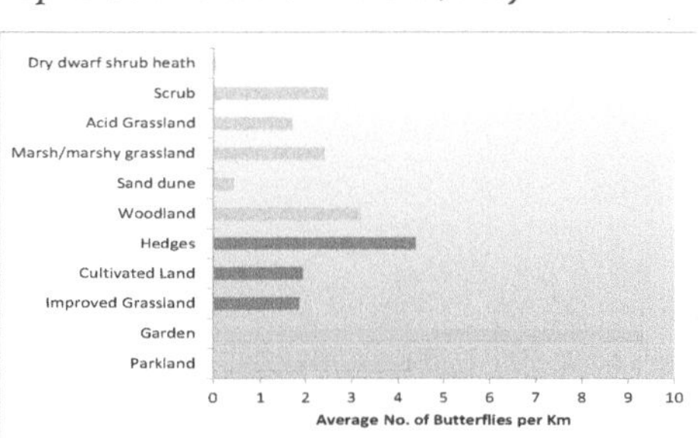

Habitat preference: No. butterflies per Km

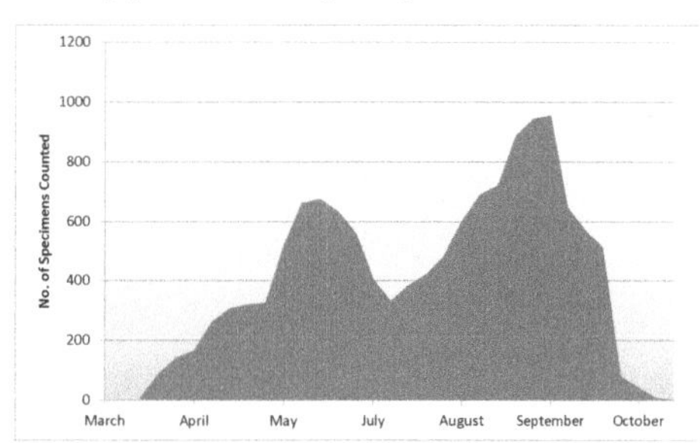

Seasonal occurrence: spring and summer

Wall Brown

Lasiommata megera

Total counted: 3,440
No. JBMS transects: 36
Wider countryside; resident

Change

-56% **UK** **+97%** **Jersey**

The Wall Brown is widespread across Jersey and has been reported from all but two of the JBMS transects. It is generally reported in low numbers except on cliff-top transects such as Les Landes and L'Oeillère and on the coastal dunes and lowland areas at St Ouen's Bay where it may be abundant.

The population trend for the Wall Brown suggests that the species is increasing sharply in Jersey but this has been distorted by just one transect, Les Landes, where the local population dramatically increased in 2011 and has remained high ever since. With this transect excluded, the Wall Brown population across other parts of the island appears to be stable. This is in contrast to the UK where it is in decline.

Records from individual JBMS transects suggest that the Wall Brown can be abundant some years and then virtually absent the next; this sudden fluctuation is unexplained.

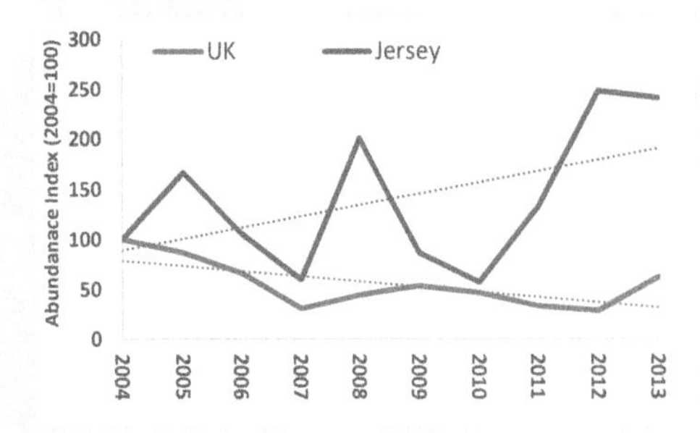

Population trend: UK versus Jersey

Butterfly and larval food plant distribution

Habitat preference: No. butterflies per Km

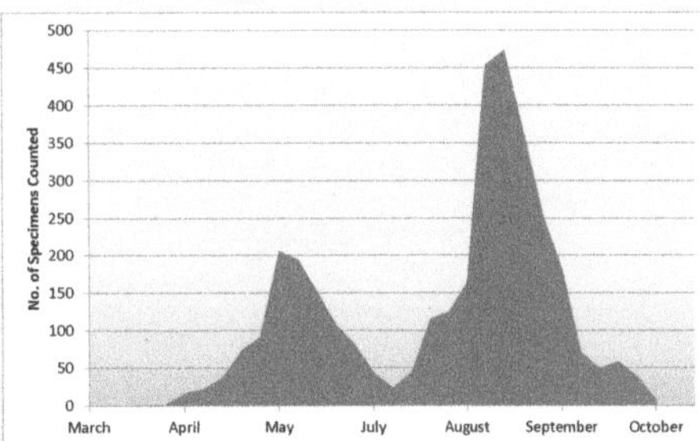

Seasonal occurrence: spring and summer

Grayling

Hipparchia semele

Total counted: 6,112
No. JBMS transects: 23
Habitat specialist; resident

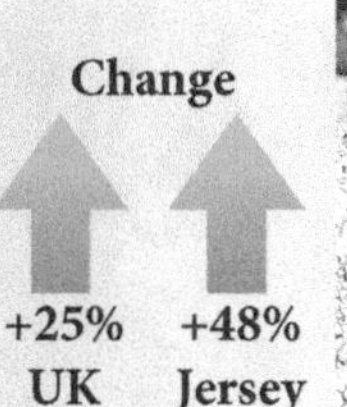

The Grayling is one of Jersey's few habitat specialists. This is reflected in its distribution which is almost entirely concentrated into two transects at Les Landes and Les Blanches Banques. Here it may be seen in large numbers.

At other JBMS transects it is rarely reported although it has been sighted on western cliff-tops such as L'Oeillère. The JBMS figures suggest that where the Grayling occurs it is doing well. However, the concentration of Jersey's breeding population into a few restricted locations makes it vulnerable to the effects of fragmentation (see Section 4.2).

As a predominantly coastal species with a preference for sunny, open locations the Grayling is suited to Jersey and its population, although variable, increased steeply during the JBMS monitoring. In order to secure its future, colonies will probably need to become established at more locations than at present.

Population trend: UK versus Jersey

Butterfly and larval food plant distribution

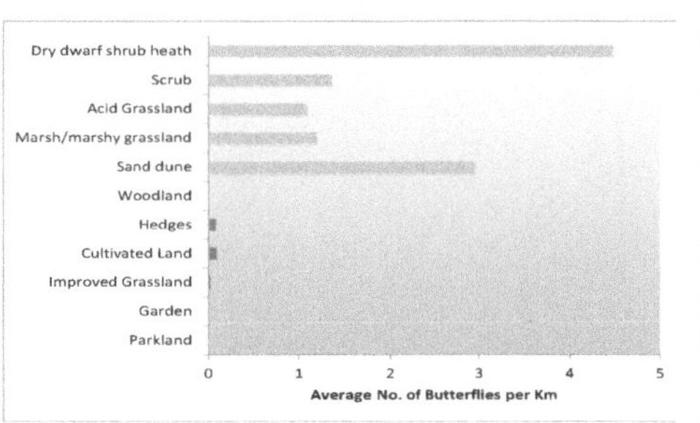

Habitat preference: No. butterflies per Km

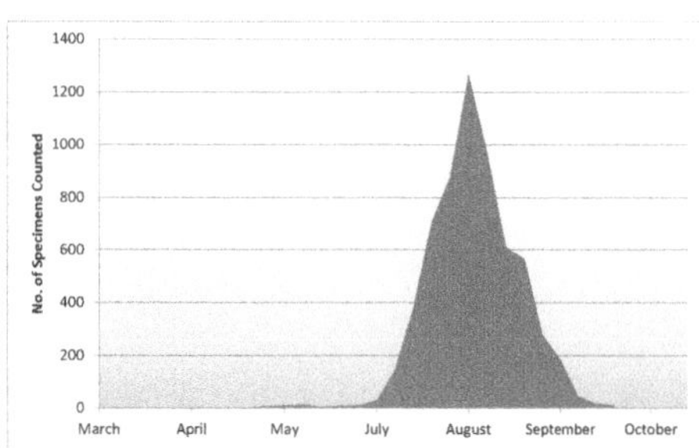

Seasonal occurrence: spring and summer

Gatekeeper

Pyronia tithonus

Total counted: 18,237
No. JBMS transects: 38
Wider countryside; resident

The Gatekeeper is Jersey's commonest and most widespread butterfly with over 18,000 reports from across all the JBMS transects. It frequents every habitat type but over half the JBMS reports are concentrated into a handful of transects on Jersey's cliff-tops and the coastal areas at St Ouen and Grouville.

The Gatekeeper's distribution in Jersey probably reflects its preference for long grass and tall vegetation along the verges of relatively open sites. The island's green lanes and field verges should suit this species but the JBMS data suggest that this species is performing poorly in these habitats.

The Gatekeeper population is declining in Jersey with much of this occurring on sites outside of semi-natural areas. The Gatekeeper can do well on verges and hedgerows and yet this does not appear to be the case in Jersey. The restoration of hedgerows and the careful management of verges could help to reverse the decline of this species.

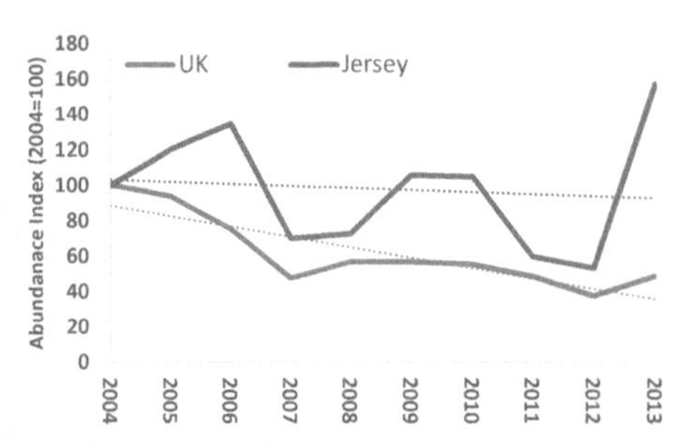

Population trend: UK versus Jersey

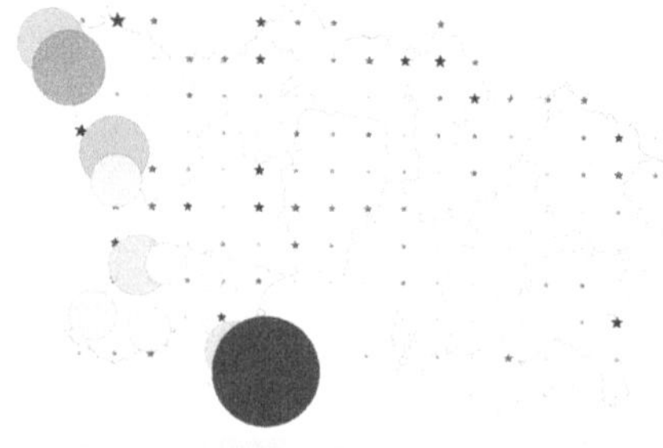

Butterfly and larval food plant distribution

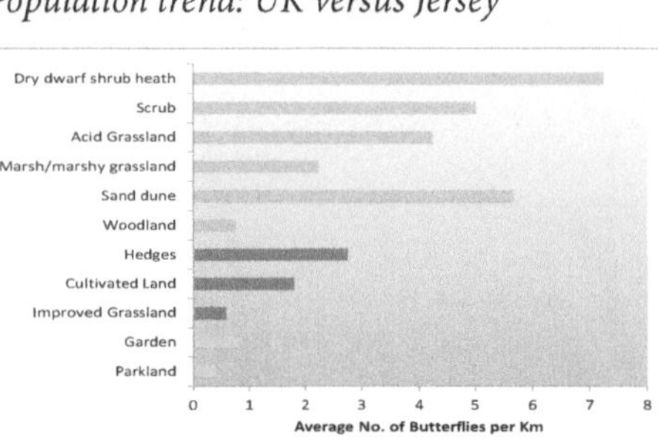

Habitat preference: No. butterflies per Km

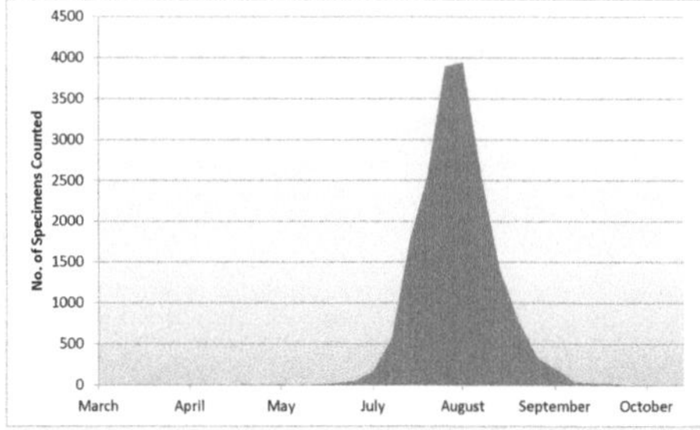

Seasonal occurrence: spring and summer

Meadow Brown

Maniola jurtina

Total counted: 13,795
No. JBMS transects: 36
Wider countryside; resident

The Meadow Brown is a widespread butterfly that has been reported from all habitat types across the island although it seems to fare less well in gardens, parks and in fields that are intensively cultivated.

The Meadow Brown population in Jersey shows a steep increase but much of this is due to its success on one agricultural transect, Field A, that was subject to an agri-environment scheme from 2007 (Section 3.2). Here Meadow Brown numbers have increased dramatically but the species also shows increases at several other JBMS transects.

As a species that frequents grass verges and hedgerows, the Meadow Brown should probably be doing better in the countryside than the JBMS data currently suggest. In the UK (where it is declining) it is suspected that the Meadow Brown does not thrive well in intensively farmed areas and this may also be the case in Jersey.

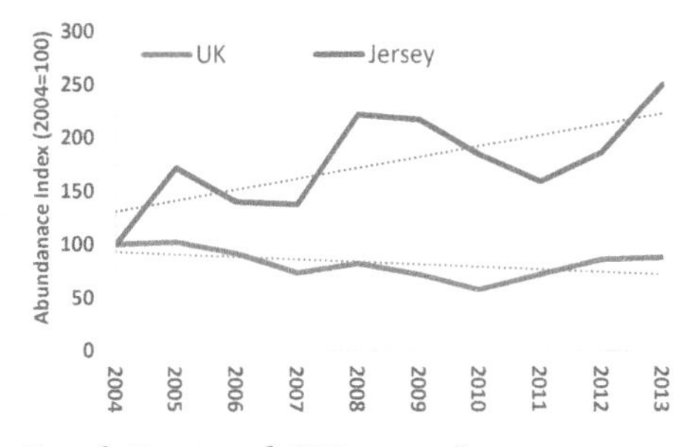

Population trend: UK versus Jersey

Butterfly and larval food plant distribution

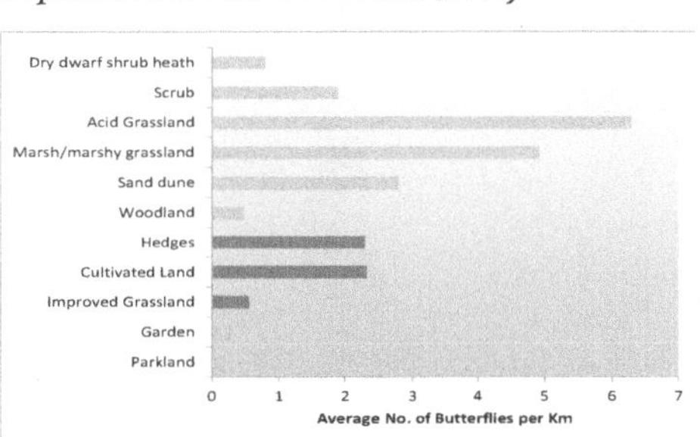

Habitat preference: No. butterflies per Km

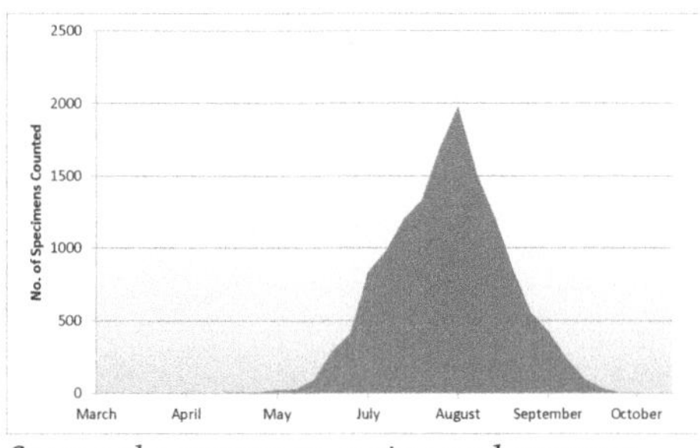

Seasonal occurrence: spring and summer

Small Heath

Coenonympha pamphilus

Total counted: 7,078
No. JBMS transects: 29
Wider countryside; resident

Change

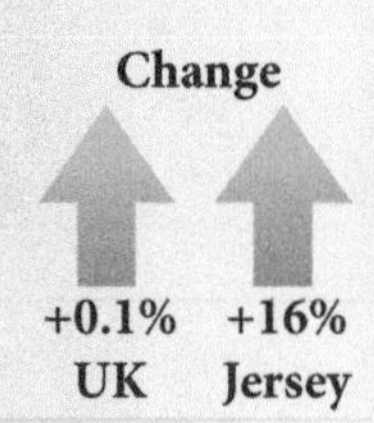

+0.1% **+16%**
UK **Jersey**

Although widely recorded in Jersey, almost all the JBMS Small Heath records come from a small handful of semi-natural transects on the west of the island but especially at Les Landes and the coastal transects at St Ouen's Bay. Smaller numbers have been reported from other cliff-tops and Grouville Golf Course but it is rarely seen outside of these areas.

The Small Heath's distribution in Jersey reflects its preference for open, grassy areas such as heaths and sand dunes. In these habitats it is doing well and the island's population is moderately increasing but the Small Heath's reliance on so few sites makes it vulnerable to habitat fragmentation (See Section 4.2).

The Small Heath's life cycle is linked to tall and medium-height grasses and, while it will probably never be common in the wider countryside, management measures along Jersey's cliffs and headlands could help to increase its distribution.

Population trend: UK versus Jersey

Butterfly and larval food plant distribution

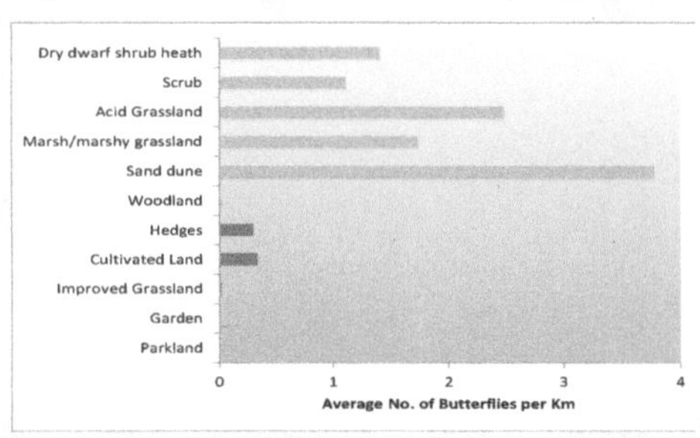

Habitat preference: No. butterflies per Km

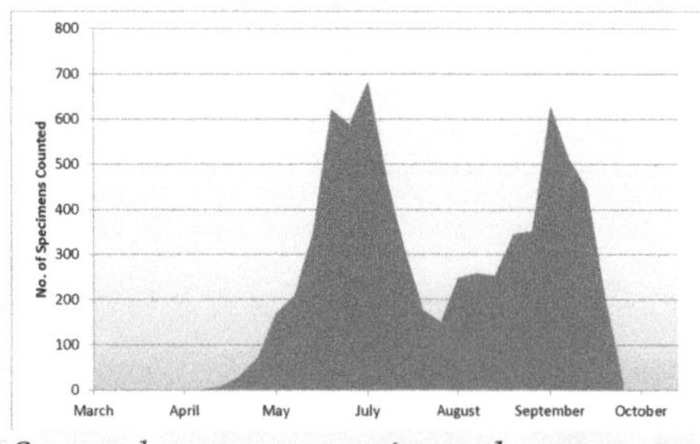

Seasonal occurrence: spring and summer

2.2 - Rare Migrants and Former Residents

The JBMS has recorded Jersey's commonest butterflies as well as several rarer species. However, Jersey's geographical proximity to continental Europe means that the island sometimes receives some rare migrants which have not been picked up by regular monitoring. In addition there are other butterfly species that were at one time resident but have since become extinct.

A supplementary list of all the Jersey butterfly species that have been recorded outside of the JBMS is provided here. If you seen any of these (or other unusual) butterflies then please photograph them and then make a report to either the Jersey Biodiversity Centre, Société Jersiaise or the Department of the Environment.

Large Chequered Skipper

Heteropterus morpheus

First observed: 1946
Last observed: 1998

Known locally by its French name, Le Miroir, this beautiful butterfly is thought to have been introduced into Jersey during the German Occupation. Although initially successful, just one colony was established at a site in Trinity. By the 1990s this site had become overgrown, with Clarke (1991) noting that 'the species is extremely vulnerable and urgent action is necessary to prevent its extinction'. A survey in July 1992 produced just nine reports and concluded that the colony was 'close to extinction' (Baker, 1992).

There have been no confirmed reports since this time and the species is presumed to be locally extinct. It has been suggested that the Large Chequered Skipper should be reintroduced to the island but the necessary investigation and planning to enable this have not yet taken place.

Grizzled Skipper

Pyrgus malvae

First observed: 1871
Last observed: 1871

Known in Jersey from a single nineteenth century record in Swiss Valley, St Saviour. It is a habitat specialist that is unlikely ever to have been resident in Jersey.

Scarce Swallowtail

Iphiclides podalirius

First observed: 1860
Last observed: 2004

CVS

A rare migrant that is rarely reported from the Channel Islands. The first Jersey sighting was circa 1860 but it was not until 1996 that it was seen again, at Victoria Tower, St Martin. It has since been seen at two other locations within the island.

Wood White

Leptidea sinapis

First observed: 1872
Last observed: 1971

CVS

A very rare butterfly with just two confirmed reports. At the turn of the twentieth century the Wood White was a common sight in Brittany and it has been speculated that this species could have been resident on Jersey prior to the advent of intensive agriculture in the nineteenth century.

Southern Brimstone

Gonepteryx cleopatra

First observed: 1986
Last observed: 1986

CCL

There is just one confirmed report of the Southern Brimstone butterfly (also called the Cleopatra). It was sighted in August 1986 on Sea Radish plants at La Pulente in St Brelade. It is very difficult to distinguish this species from the Brimstone (the main difference is the shape of the wings when closed) so, although undoubtedly rare in Jersey, it may have been under-reported.

Black-veined White
Aporia crataegi

First observed: 1862
Last observed: 1862

Although recorded in a list of Jersey butterflies published in 1862, there have been no further reports of the Black-veined White and so it is at present an unconfirmed species for the island. This species went extinct in the UK in the 1920s but was historically common in Europe and so it is possible that it was on the island in the early nineteenth century.

Bath White
Pontia daplidice

First observed: 1834
Last observed: 1945

Listed as abundant in Jersey in 1834 and as being locally common during most of the nineteenth century, by 1910 the Bath White was extinct in Jersey. Its caterpillars are intolerant of the cold and it is possible that the species was lost during one of the severe winters in the 1890s. An influx of migrant specimens from Europe occurred in 1945 but it has not been seen since.

Large Copper
Lycaena dispar

First observed: 1862
Last observed: 1862

Listed as a Jersey butterfly in 1862, the Large Copper has no confirmed records from the island. The Large Copper is a specialist species with a preference for fens, marshland and wet meadows, none of which are extensive in Jersey. However, the south of the island did have some marshland in the nineteenth century so its former presence here is at least conceivable.

Short-tailed Blue

Everes argiades

First observed: 1942
Last observed: 1944

Just two records were made, both during the German Occupation of 1940 to 45 (as were records for a number of other rare butterflies hinting that conditions in occupied Europe may have been favourable to butterflies). These specimens were almost certainly migrants from Europe.

Mazarine Blue

Cyaniris semiargus

First observed: 1942
Last observed: 1946 (2002?)

As with the preceding species, several specimens of the Mazarine Blue were observed in the east of Jersey during and just after the German Occupation. It has been extinct in the UK for decades but is still found on mainland Europe. Its preference for uncultivated flower-rich meadows makes it unlikely that the Blue Mazarine could establish itself in Jersey although there is a reliable sighting of a single specimen in Grouville during 2002.

Purple Emperor

Apatura iris

First observed: 1940
Last observed: 2014

An exceptionally rare and only one has been seen alive in Jersey. One dead specimen was found in 1940 from Noirmont Manor headland and a second moribund specimen was found on St Catherine's slipway in 2014. Both specimens will almost certainly have been blown across from France.

Camberwell Beauty

Nymphalis antiopa

First observed: 1860
Last observed: 1986

There are rare but sporadic reports in Jersey of this strong-flying butterfly, all of which represent migrant specimens. It is unmistakable and nomadic with the few Jersey reports most often coming from urban and domestic settings including one which flew into a school classroom.

Pearl-bordered Fritillary

Boloria euphrosyne

First observed: 1947
Last observed: 1947

Known from a single specimen taken near Bouley Bay, Trinity, in 1947. Although widespread in Europe, the Pearl-bordered Fritillary's preference for open woodland habitats, which are limited on Jersey, makes it an unlikely resident for the island. The single record probably represents a migrant from continental Europe.

Dark Green Fritillary

Argynnis aglaja

First observed: 1872
Last observed: 1951

Recorded regularly but sporadically on Jersey, the Dark Green Fritillary has been locally extinct since the 1950s. The Jersey records probably represent migrant individuals and there is no firm evidence that colonies have ever existed. A thriving colony is established in Sark and is not uncommon on Herm.

Silver-washed Fritillary

Argynnis paphia

First observed: 1917
Last observed: 1946

There are just three reports of the Silver-washed Fritillary, two of which probably represent migrant specimens. The third sighting was of three adults seen near St Catherine's woods in 1946 which may represent a short-lived colony in the area. As a frequenter of hedgerows, scrub and abandoned land, the Silver-washed Fritillary is widespread and tolerably common in the UK and France. It probably could establish colonies on the island albeit locally.

Glanville Fritillary

Melitaea cinxia

First observed: 1860
Last observed: 1992

The Glanville Fritillary was once widespread in Jersey and was regularly described as being common or abundant. Following the Second World War several of its key habitats and hotspots (such as Bellozanne Valley) became industrialised or fragmented and by the 1980s it was resident only at Les Landes.

The stormy winter of 1988 scorched the coastal heathland in St Ouen after which there were few Glanville Fritillary reports, the last of which was in 1992. The history of the Glanville Fritillary on Jersey would seem to be an example as to how habitat fragmentation can render a once abundant and widespread species locally extinct over a short time period.

Marbled White

Melanargia galathea

First observed: 1999
Last observed: 1999

There is one record of a single Marbled White being in St Catherine's Woods from 1999. It is not known from the other Channel Islands but is locally common in Europe and so presumably this was a migrant from the continent.

Monarch

Danaus plexippus

First observed: 1989
Last observed: 2004

CVS

The Monarch is famous for its trans-American migrations and individuals, which might well have originated in the Canary island, where it is well established, sometimes find their way to Europe. The three Jersey reports of Monarchs may represent migrant individuals or possibly specimens that have escaped or been released after being captive bred (see next page).

Doubtful Reports

The following species were recorded during the JBMS but are thought to represent misidentifications.

Dingy Skipper (*Erynnis tages*).

Reported twice in quick succession from one JBMS transect. With no other Jersey records at all it is felt that these were misidentifications.

Small Skipper (*Thymelicus sylvestris*).

This species is very difficult to tell apart from the Essex Skipper in the field. There were several Small Skipper reports as part of the JBMS but we have adopted the view of Roger Long (pers. comm.) who reports that all preserved local specimens labelled as Small Skippers have proved to be misidentifications of the Essex Skipper.

Large Heath (*Coenonympha tullia*).

There was one JBMS report of a Large Heath. This is a boreal species that is rarely recorded south of Scotland. It is very unlikely to have been seen in Jersey.

It is worth noting that Long (2009) lists the following species as having doubtful records from the nineteenth century:

- **Silver-spotted Skipper** (*Hesperia comma*)
- **Apollo** (*Parnassius apollo*)
- **Silver-studded Blue** (*Plebejus argus*)
- **Weaver's Fritillary** (*Boloria dia*)

2.3 - Captive Releases

In recent years there has been a fashion for 'butterfly kits' which can be bought cheaply on the internet. The kits provide the necessary equipment and larvae to permit the raising of exotic butterflies. However, the cages that they come in are often confined and once the adult butterfly emerges, it is often seen as being kinder to release the animal into the wild than keep it in a restricted space.

Similarly, there is a growing trend for releasing clouds of captive bred butterflies as part of wedding ceremonies ('butterfly confetti'). With kits and 'confetti' the released butterflies tend to be large, tropical species that may survive for a short while in Jersey's climate before succumbing to the cold or predators.

The release of non-native species into the wild on Jersey is discouraged (and in some cases illegal) and doing so causes distress to the released animals which are not suited to the island's weather or food plants. It also risks introducing new diseases into the native butterfly population.

This practice has also led to confusion among Jersey naturalists who have made three recent reports of released butterflies in Jersey. One of these was initially thought to represent the first British record of a continental species and caused much excitement until its true origin was established.

Released butterflies tend to be large tropical species such as the Julia (*Dryas iulia*) and Plain Tiger (*Danaus chrysippus*) but species such as the Monarch (*Danaus plexippus*) may also be used. The Department of the Environment is strongly against the practice of releasing captive bred butterflies and requests that before doing so people think of the wellbeing of the local environment and the animals concerned.

Julia Butterfly - *Dryas iulia*

Plain Tiger - *Danaus chrysippus*

Please do not release captive bred butterflies into the wild.

- Section Three -

Butterfly Population Trends 2004 to 2013

3.1 - Jersey versus UK Butterfly Trends

Because monitoring in Jersey has been carried out in an identical manner to that in the UK, the JBMS ten-year dataset can be directly compared with figures from the UKBMS over the same time period.

A comparative analysis between the JBMS and UKBMS datasets was kindly performed by Dr Tom Brereton of Butterfly Conservation. The results for the 24 commonest JBMS species monitored between 2004 and 2013 are presented in Table 3. To accompany this analysis Dr Brereton has written a short report which is presented below.

Individual Species' Trends

On Jersey ten of 24 species (42%) assessed show a negative change, while fourteen species (58%) show an increase. In the UK the trends for the same 24 species show that 20 (83%) experienced a negative change, one showed no overall change, whilst three species (13%) showed an increase. For 18 of the 24 species (75%), the percentage change in abundance over the decade was more favourable on Jersey compared with the UK (Table 3).

Analysis of covariance (Ancova) was used to test for trend differences between Jersey and the UK. Three species had significantly more favourable trends on Jersey compared with the UK. These are: the Green Hairstreak, the Meadow Brown and the Wall Brown. Another two additional species were close to having more favourable trends: Essex Skipper and Small Copper. In contrast, Small White and Swallowtail had less favourable trends on Jersey compared with the UK.

The Small Copper, Green Hairstreak and Wall Brown may possibly be doing better on Jersey compared with the UK. Given that in Jersey the coast is never very far away, this success may reflect the greater stability and resilience of butterflies in coastal areas. In particular, the Wall Brown is known to be declining in inland areas of lowland England but has maintained its range in coastal areas.

The decline of the Swallowtail is concerning and may be attributable to habitat deterioration which could be ameliorated by enhanced management. A less favourable trend for the Small White on Jersey might suggest an impact from agricultural intensification.

Migration pulses were higher on Jersey than the UK average. This was expected given that Jersey is located further to the south and therefore a nearer port of call for migrants from southern and central Europe.

Species	Change UK	Change Jersey	Change difference	Trend difference
Brimstone	-30%	-25%	5%	None
Brown Argus	-34%	-66%	-32%	None
Clouded Yellow	-88%	-91%	-3%	None
Comma	-25%	8%	33%	None
Common Blue	-27%	28%	55%	None
Essex Skipper	-81%	24%	105%	Near better Jersey
Gatekeeper	-61%	-22%	39%	None
Grayling	25%	48%	23%	None
Green Hairstreak	-40%	458%	498%	Better Jersey
Green-veined White	26%	17%	-9%	None
Holly Blue	-60%	11%	71%	None
Large Skipper	-16%	108%	124%	None
Large White	-1%	-21%	-20%	None
Meadow Brown	-25%	90%	115%	Better Jersey
Painted Lady	-89%	-55%	34%	None
Peacock	-6%	153%	159%	None
Red Admiral	-14%	1%	15%	None
Small Copper	-19%	51%	70%	Near better Jersey
Small Heath	0%	16%	16%	None
Small Tortoiseshell	-28%	95%	123%	None
Small White	8%	-67%	-75%	Worse Jersey
Speckled Wood	-9%	-8%	1%	None
Swallowtail	-13%	-95%	-82%	Worse Jersey
Wall Brown	-56%	97%	153%	Better Jersey
All species combined	-29%	-14%	15%	None

Table 3. Changes in butterfly species' abundance 2004-2013: Jersey versus the UK as a whole. (Source: Butterfly Conservation)

All Species Combined

The grouped measures of butterfly abundance for 24 species for the ten-year period 2004-13 indicated that the island's abundance apparently declined by 14% on Jersey and by 29% across the UK (Fig. 3; Table 3).

Neither of these declining trends (assessed by ordinary least squares regression) were statistically significant. Similarly, the apparent difference in the rate of decline between Jersey and the UK (assessed by ANCOVA) was not statistically significant.

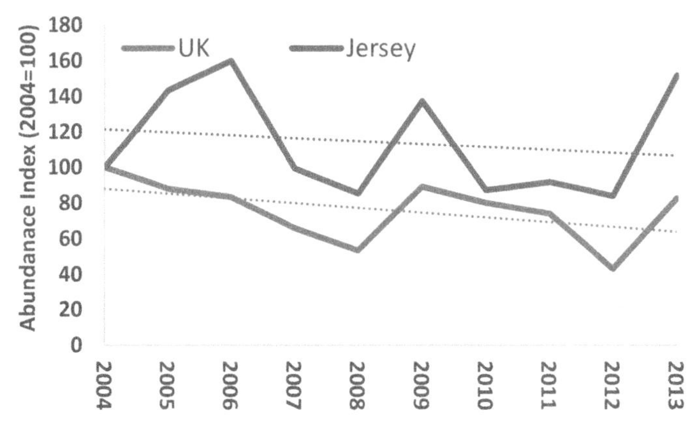

Figure 3. Trends in composite butterfly abundance (n = 24 species) on Jersey compared with the UK as a whole. Dotted lines represent underlying linear trends.

3.2 - Habitats and Butterfly Trends

As well as counting butterfly species, the JBMS also recorded the habitats of the individual transect sections walked by our volunteers (see Section 1.2). The transect sections were assessed annually and any changes in habitat, land use (e.g. the planting of new crops) or management (e.g. mowing, cutting or spraying) were noted by volunteers during their weekly walks.

The JBMS section habitats have been classified using both the European EUNIS scheme (as used by Butterfly Conservation) and the JNCC's Phase 1 classification system. The latter classification was applied retrospectively so that the ten-year analysis could utilise the results from a whole island Phase 1 survey undertaken in 2011.

Broad Transect Classification

The transects monitored during the JBMS were grouped into three broad categories based on their principal land use and management practices. These are:

Semi-natural

Those habitats that retain many natural features but are modified by human influence. (Natural sites are those that have no human modification at all; Jersey has none of these.)

Agricultural

Those habitats that are utilised for farming practices such as growing crops and grazing. This includes conservation farming schemes.

Urban

Highly modified and managed habitats which includes parklands, gardens and amenity areas such as cemeteries, playing fields, etc.

A fourth category covering conservation farming schemes was created at the start of the JBMS but these schemes were mostly short-lived (see Section 3.3) and have since been included within the agricultural category.

Figure 4 displays the JBMS butterfly trends between 2004 and 2013 for the three broad transect categories. This suggests that butterfly populations in semi-natural habitats are increasing while those in agricultural habitats are stable and those in urban habitats are steeply declining. However, these results may be broken down further by analysing the Phase 1 habitats that constitute the three broad transect categories.

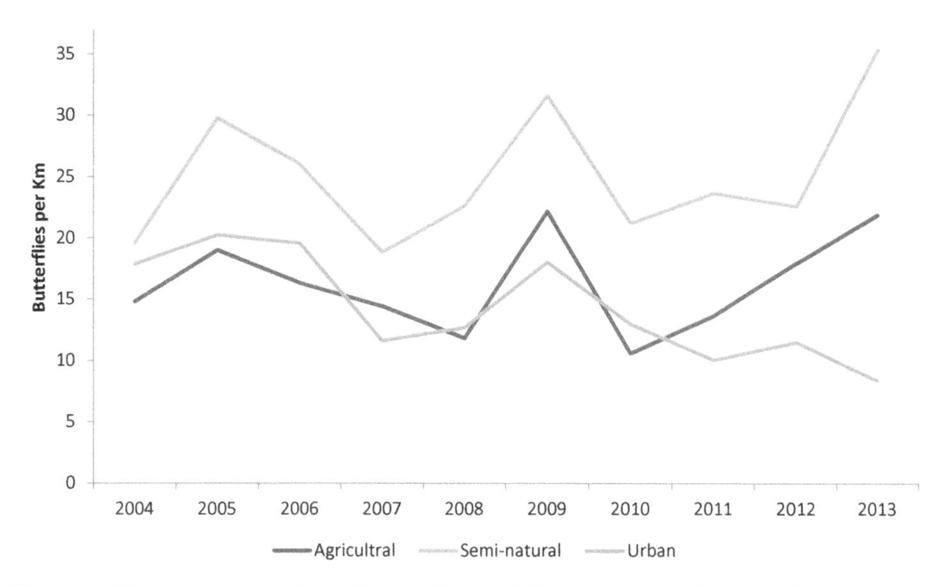

Figure 4. The average number of butterflies per kilometre across the three broad transect categories.

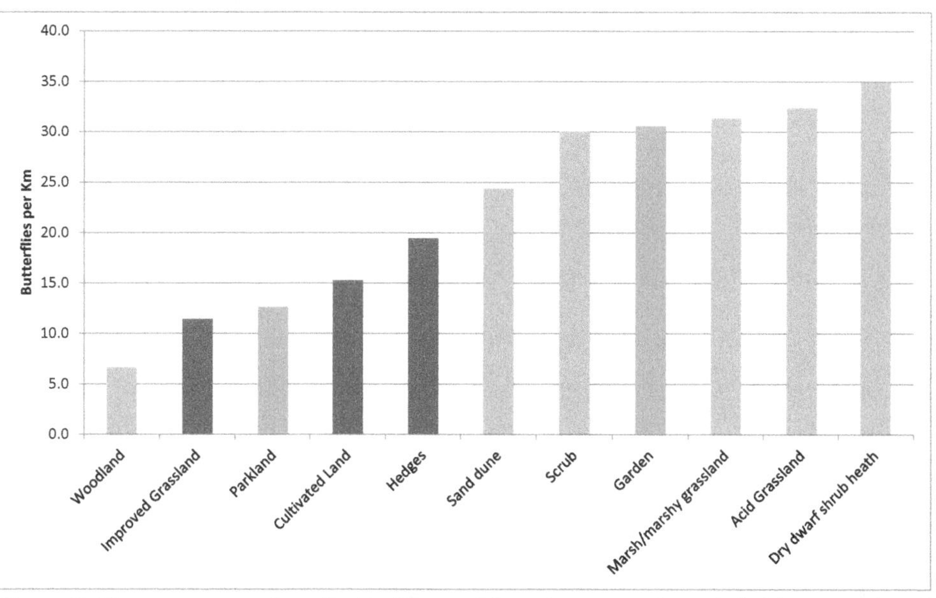

Figure 5. The average number of butterflies per kilometre across the main habitats monitored by the JBMS (2004 to 2013). Green = semi-natural habitats; blue = agricultural; orange = urban.

Figure 5 shows the average abundance of butterflies within the Phase 1 habitats that were monitored by the JBMS. This suggests that there is considerable variation in butterfly tolerance between habitats. To quantify this better, the population trends within the Phase 1 habitats were examined in more detail.

Individual Habitat Accounts

The JBMS records were analysed with a view to understanding the health of butterfly populations within the principal monitored Phase 1 habitat types. The results suggest that there are marked differences in the abundance, composition and population trends between these habitats.

This section provides individual accounts of the main Phase 1 habitats that were monitored as part of the JBMS. As well as a general discussion, the following statistical information is provided.

Distance Surveyed. The collective transect distance surveyed between 2004 and 2013. This was calculated using the length of transect sections and the number of times that it had been walked by volunteers during the ten-year monitoring period.

Total Butterflies Counted. The number of recorded butterfly reports for each habitat 2004 to 2013. As above, this could be calculated using the count data for the habitats recorded for each transect section.

Butterflies per Kilometre. The average number of butterfly reports per kilometre between 2004 and 2013. This provides a guide as to the abundance of butterflies within the habitat. The more butterflies per kilometre, the greater their abundance within the habitat.

Total Diversity. The number of species recorded from this habitat. This includes all species recorded and does not differentiate those that were common from those that are rare.

Principal Species. A list of the commonest species recorded from each habitat between 2004 and 2013. The figure in brackets represents the percentage of all butterfly reports for the habitat.

Population Trend. This line graph displays the average number of butterflies observed per kilometre each year. This offers a visual guide to changes in the annual butterfly population. The black line represents the linear regression trend.

It should be noted that some of the habitat trends reflect the effect of the large Painted Lady influx of 2009 (and, to a lesser degree, 2005), the butterfly-poor springs and summers of 2011 and 2012 (see Section 3.6) and a marked population increase in the Green Hairstreak, Grayling, Small Tortoiseshell and Large Skipper in 2013.

Phase 1 Habitat Map. The extent of the habitat on Jersey in the 2011 Phase 1 survey undertaken by the Department of the Environment.

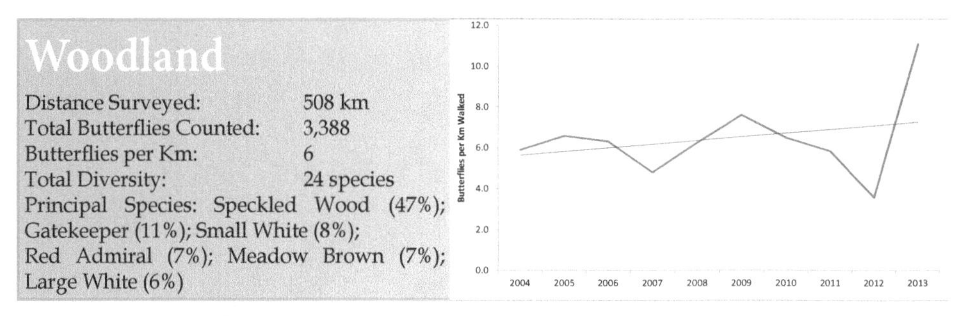

Jersey's woodland areas are mostly broad-leaved, of small extent and restricted to valleys. They tend to be quite young as many trees were cut down and used for fuel during the German Occupation.

Other than some specialist species, woodland habitats do not generally suit butterflies and, of all the Jersey habitats monitored, this was the poorest for butterfly reports in terms of abundance. However, the ten-year JBMS trend suggests that the butterfly population in Jersey's woodlands is slightly increasing although this may be influenced by an unusually high count of Speckled Woods in 2013.

Jersey's woodlands may not be the best general butterfly areas but they do attract rare resident and migrant butterflies such as the White Admiral. The relative stability of Jersey's woodland habitats may favour the establishment of rarer species, especially on sites that are managed.

Jersey's woodland habitat is mostly associated with the valleys that generally flow south across the centre of the island.

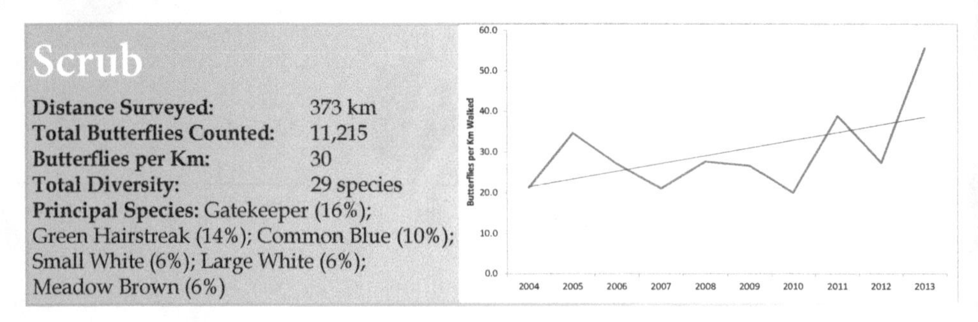

Scrub type habitat (defined as a mix of native shrubs with scattered trees) is mainly located in semi-natural areas on the west and very north of the island. They are particularly common on coastal slopes and cliff-tops but becomes scarcer and more fragmented inland and on the east of the island.

Jersey's scrub habitat is generally favourable for butterflies and JBMS records suggests that it supports a diverse range of species, including habitat specialists such as the Green Hairstreak and Grayling. The JBMS data suggest that scrub habitats are an important habitat for Jersey's butterflies and that their populations are steeply increasing.

However, a lack of connectivity between areas of scrub is possibly leaving butterfly populations isolated and which may make them vulnerable to the sort of step-wise decline already observed in some species such as the Glanville Fritillary.

Jersey's scrub habitat is predominantly located on western and northern cliff-tops. They tend to be fragmented and discontinuous.

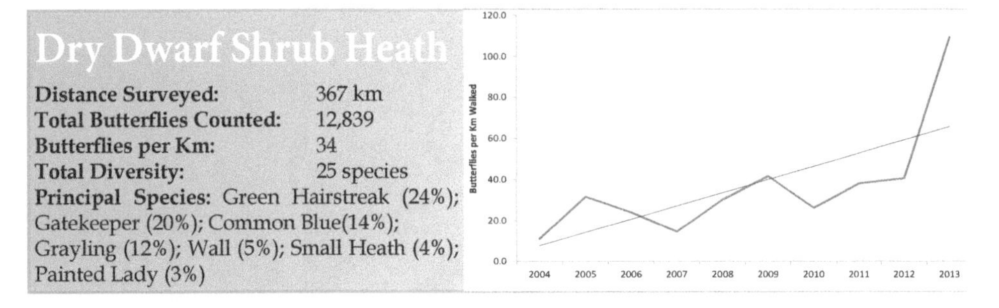

Dry dwarf shrub heath is an open habitat consisting of heather and species such as gorse. In Jersey it is associated with coastal escarpments and cliff tops, especially in the north-west of the island. During the ten-year JBMS the largest area of dry dwarf shrub heath monitored was in the Les Landes area so the JBMS results for this habitat may not be representative of other similar island sites.

The dry dwarf shrub heath habitat has the greatest abundance of butterflies measured in the whole JBMS and also a high diversity of species. The butterfly population steeply increased across the decade and includes habitat specialist species such as the Green Hairstreak and Grayling, making this the most successful butterfly monitored area in Jersey. It would be useful to know whether this success is just related to the Les Landes site or whether it applies to dry dwarf shrub heath elsewhere in the island. This habitat needs careful monitoring to ensure it remains in good condition.

Dry dwarf shrub heath is restricted to a few of Jersey's cliff-tops. The largest expanse is at Les Landes; at other locations it tends to be fragmented.

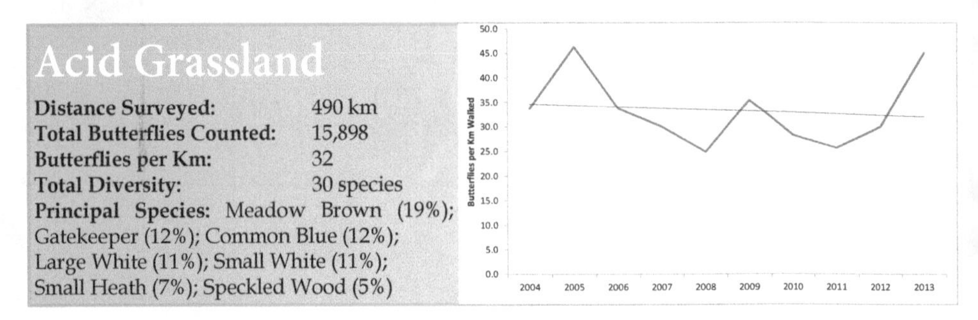

Acid grasslands form on nutrient-poor soils and may contain a mix of tall and medium grasses, heather, gorses and other plants. The habitat most commonly occurs on the top of escarpments and cliff sites in the west and south-west of the island although smaller patches may be found close to the north and east coasts.

This is a good habitat for butterflies which may be due to a predominance of grass species (on which many Jersey butterflies preferentially feed and pupate) and connectivity to scrub or dry dwarf shrub heath habitats which often occur in the same general area.

Butterfly populations are stable or possibly slightly declining on Jersey's acid grasslands. This habitat is vulnerable to change through over or under management which can cause grass species to become overwhelmed by other vegetation or to be kept too short for butterflies to utilise (see also Section 3.5).

Acid grassland habitat is mostly associated with cliffs and coastal escarpments in the west of Jersey.

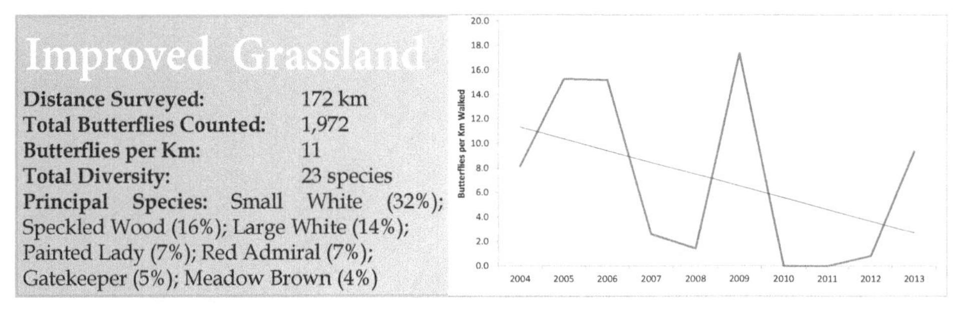

Improved grassland covers grassy meadows and pastures that are intensively grazed or which are managed using agricultural chemicals, manure or slurry. It is an agricultural habitat that is common across Jersey especially inland from the coast.

The JBMS data suggest that butterflies do not favour improved grassland areas with a majority of the species seen there being strong-flying species which could have migrated in from elsewhere. Evidence from the JBMS suggests that, in contrast to cultivated land, butterfly species are regularly recorded in the field centre as well as the margins. This could be due to the presence of flowers (such as thistles) in the middle of fields.

Butterfly populations on improved grassland seem to be steeply declining, possibly because many of the butterflies recorded there are vagrant rather than resident. Given that this habitat covers a large area of the island, this could present an issue for Jersey's insect biodiversity.

Improved grassland habitat may occur anywhere on the island but is generally a feature of agricultural areas inland from the coast.

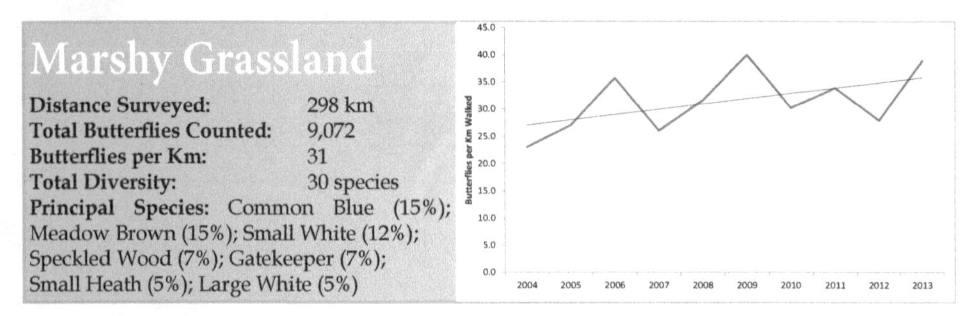

Marshy Grassland

Distance Surveyed: 298 km
Total Butterflies Counted: 9,072
Butterflies per Km: 31
Total Diversity: 30 species
Principal Species: Common Blue (15%); Meadow Brown (15%); Small White (12%); Speckled Wood (7%); Gatekeeper (7%); Small Heath (5%); Large White (5%)

Prior to the twentieth century Jersey had a sizeable area of marshy ground in the south and east of the island but this was gradually built over as housing and other infrastructure expanded along the coastal plain. Relatively few marshy sites remain in Jersey and most are associated with semi-natural areas in inland valleys or coastal sites such as St Ouen's Pond.

Jersey's marshes and marshy grassland areas are good habitats for butterflies both in terms of diversity and abundance. The JBMS data suggest that populations on this habitat are moderately increasing and that there is a good selection of resident species. However, Jersey mostly lacks the habitat specialist species seen in UK marshy sites; this may be a factor of their small size and often isolated occurrence.

Relatively few JBMS transects contain marshy ground and it is possible some of the island's other wet meadow areas (such as St Peter's Valley) will prove to be good areas for butterflies.

Jersey's marshy grassland habitat is associated with inland valleys and coastal dune sites. Many former marshy areas have been lost to housing development.

Sand Dunes

Distance Surveyed: 981 km
Total Butterflies Counted: 23,954
Butterflies per Km: 24
Total Diversity: 29 species
Principal Species: Gatekeeper (23%); Common Blue (15%); Small Heath (15%); Grayling (12%); Meadow Brown (11%); Painted Lady (3%); Large White (3%)

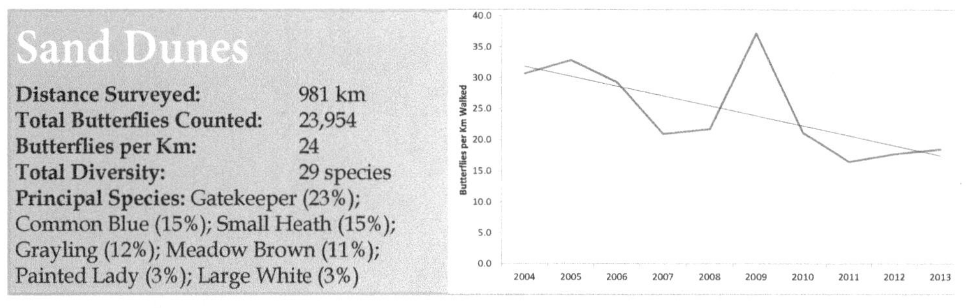

Jersey's principal sand dune area lies on the coastal plain at St Ouen's Bay although smaller patches do exist elsewhere in Jersey. These smaller areas are often highly modified and, with the exception of Grouville Golf Course, were not monitored by the JBMS.

The St Ouen's Bay sand dune habitat is rich in plant species and good for butterflies in terms of their diversity and abundance. However, the JBMS data suggest that butterfly populations are decreasing there which, given the size and importance of the dunes for butterflies (as well the island's general biodiversity), is of great concern.

The cause of this decline has not been fully established but it is possible that some habitats are have become overgrown in recent years or that there is pressure from the large numbers of people using the dunes. An additional JBMS transect has been established to help quantify the problem and the matter will be the subject of further investigation.

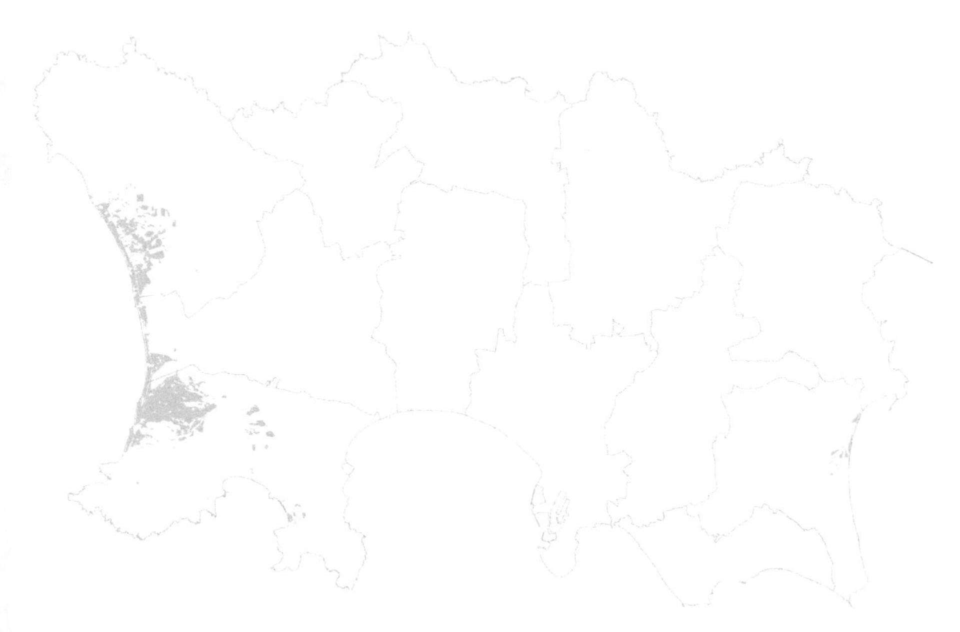

Sand dunes form the largest areas of continuous semi-natural habitat in Jersey. They are predominantly a coastal feature in the west of the island.

Cultivated Land

Distance Surveyed: 1,405 km
Total Butterflies Counted: 21,420
Butterflies per Km: 15
Total Diversity: 29 species
Principal Species: Small White (20%); Meadow Brown (15%); Large White (14%); Speckled Wood (12%); Gatekeeper (11%); Painted Lady (7%); Common Blue (4%)

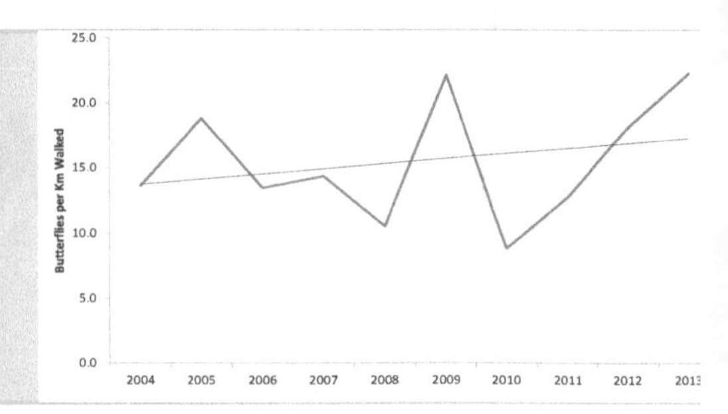

Cultivated land habitats are almost exclusively agricultural in origin and include sites which have been ploughed or on which agricultural crops (including organic crops) or monospecific grasses are grown.

Cultivated land on Jersey is not a favourable habitat for butterflies with abundance and diversity both being low. Many of the principal species are strong fliers which may have migrated in from elsewhere. Resident species are generally recorded along the field margins and not in the centre, highlighting the importance of verges to wildlife in Jersey.

The JBMS data suggests that butterfly populations on cultivated sites are highly variable from year to year (probably because some of its principal species are vagrants) but that the overall trend is either stable or slightly increasing. For an analysis of conservation farming schemes see Section 3.3.

Cultivated land dominates the agricultural landscape inland from the coastal strip. It is where the island's potatoes and other commercial crops are grown.

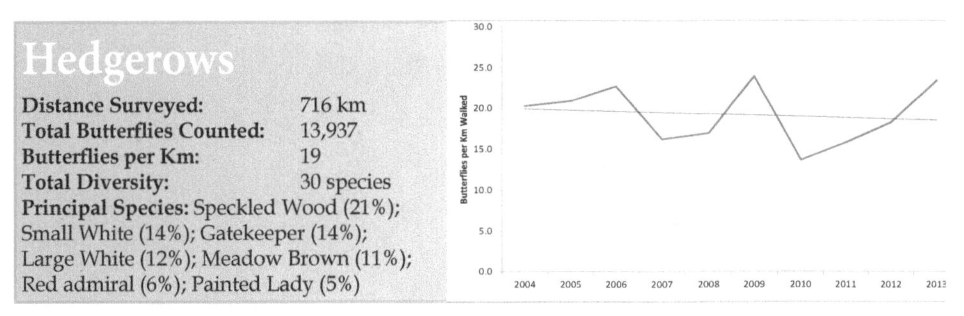

Hedgerow habitats are important in Jersey, especially in agricultural areas where, together with verges, they can provide corridors of trees, grasses and wild flowers that can be utilised by a wide range of animals from birds to bees. It should be noted that no attempt was made to grade the health status of the JBMS hedgerows.

Hedgerows are often associated with field margins and so their butterflies are similar to those recorded on cultivated land. Jersey's hedgerow butterflies are all generalist species and the ten-year trend is slightly declining. Many of Jersey's hedgerows are in a poor state of health and are probably not good habitats for insects and other wildlife. Also, the mechanical cutting of hedgerow trees during the spring and summer months does not favour wildlife (see Section 4.2). The restoration and maintenance of Jersey's hedgerows is desirable for many different reasons including the enhancement of Jersey's butterfly species.

Hedgerows are mostly associated with the margins of fields and hence have a pattern that is similar to that of cultivated land and improved grassland.

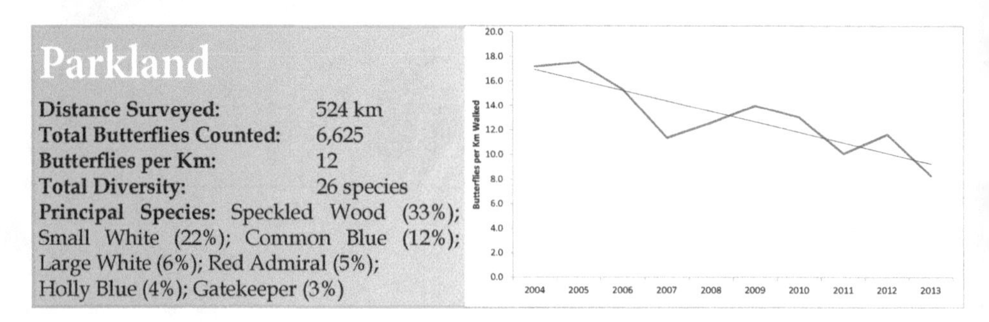

Parkland

Distance Surveyed: 524 km
Total Butterflies Counted: 6,625
Butterflies per Km: 12
Total Diversity: 26 species
Principal Species: Speckled Wood (33%); Small White (22%); Common Blue (12%); Large White (6%); Red Admiral (5%); Holly Blue (4%); Gatekeeper (3%)

Jersey's parks and managed community gardens are mostly located within, or adjacent to, built-up areas. Those monitored as part of the JBMS are mostly in or around St Helier and include Howard Davis Park, West Park, South Hill Park and Green Street Cemetery.

Managed parklands are a poor habitat for butterflies in terms of their number and diversity and they have performed consistently worse than the domestic gardens monitored by JBMS. However, a profusion of summer flowers can attract some species such as Red Admirals and Small Tortoiseshells as well as rarer ones such as the Queen of Spain Fritillary.

The JBMS data suggest that Jersey's parkland butterfly populations are steeply declining. This decline could be reversed by managing small areas of park for the benefit of pollinating insects by establishing butterfly-friendly plants and leaving some areas to grow wild grasses and nettles.

Parkland and ornamental gardens may occur anywhere in the island but are most often associated with built up areas and the grounds of large residences.

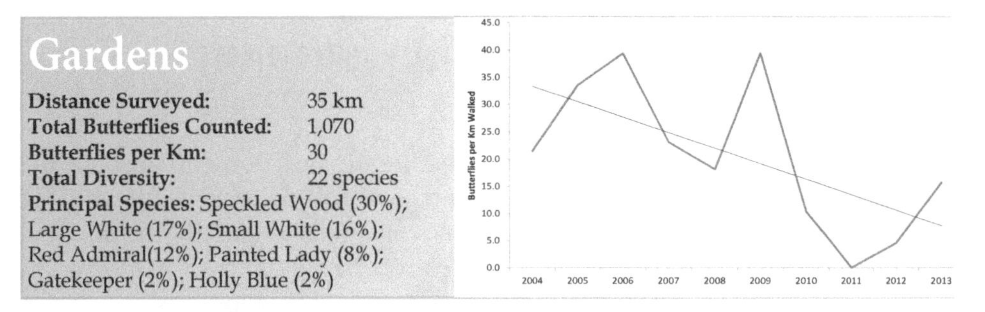

Gardens are not a specific habitat with the Phase 1 classification scheme and in the context of the JBMS this category was created to cover conventional domestic plots with flower beds, ornamental features and lawns.

Not many gardens were included within the JBMS and so these results may not be generally representative of this habitat. However, the available data suggest that Jersey's gardens are remarkably similar to parkland habitats in terms of their species and population decrease. An increase in 2013 may be due to a sharp recovery in Jersey's Small Tortoiseshell population.

Gardens are good places for a few strong-flying species but overall diversity is generally low. As with parkland, more general awareness of butterfly friendly plants and management techniques amongst householders could lead to an improvement in garden butterfly numbers.

Domestic gardens are spread across Jersey but are particularly concentrated along the south coast and on the outskirts of built up areas.

3.3 - Conservation Farming

As a southerly island with a temperate climate and rich soil, agriculture forms an important part of Jersey's land use, culture and economy. The history of the island's agriculture has changed markedly from cider orchards in the eighteenth century through to wheat and cattle in the nineteenth, then a post-war boom in tomatoes and then potatoes.

Jersey's modern agricultural landscape is dominated by potato/ brassica farming and improved grassland areas for grazing. The combined agricultural sector represents the largest single use of land with much of Jersey's interior consisting of a patchwork of small farms, lanes and boundary features that characterise the island's rural landscape.

The JBMS data suggest that agricultural habitats are poorer for butterflies in terms of abundance and diversity than semi-natural ones (see Section 3.2). These results mimic European studies which suggest that pollinating insect populations are declining in intensively farmed areas (van Swaay, 2014; Fox *et al*, 2006). In Europe attempts have been made to mitigate the effect of intensive agriculture on biodiversity through conservation farming initiatives, such as planting bird and insect friendly crops and organically grown crops, to see if they may be beneficial to wildlife.

Conservation farming schemes also operate on Jersey and one of the founding objectives of the JBMS was to see what effect these initiatives might have on the local butterfly population. Two types of conservation farming scheme were monitored by the JBMS: agri-environment schemes involving conservation crops and management, and farms that have adopted organic farming techniques.

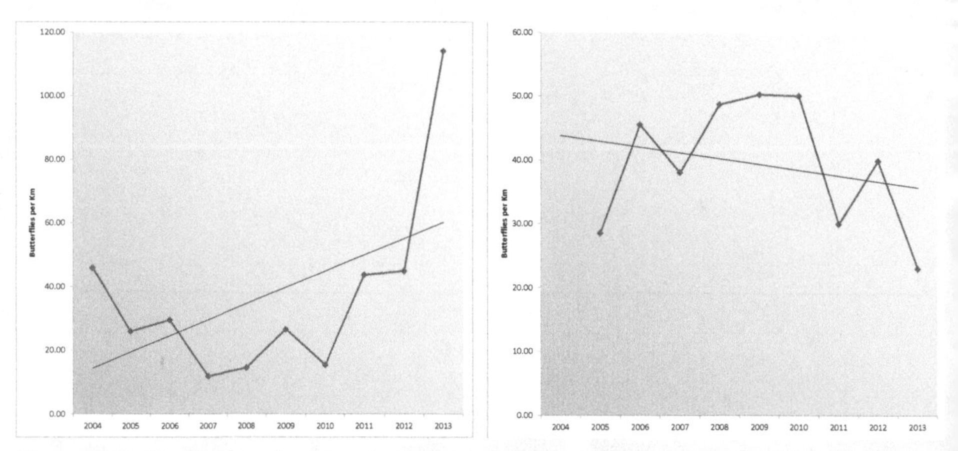

Figure 6. Butterfly abundance on two agri-environment transects. Left = Field A (agri-environment from 2007-2013). Right = Field G (agri-environment 2006-2010).

Agri-environment Schemes

The JBMS included three transects that implemented agri-environment schemes for some of the time that they were monitored. The schemes managed land for wildlife via bird/insect-friendly planting, grassland management and a prohibition on the use of pesticides and herbicides.

These schemes were sponsored through the States of Jersey Countryside Renewal Scheme (CRS) which operated between 2005 and 2012. All but one of the agri-environment schemes stopped operating when their CRS funding ceased. Consequently the JBMS does not have ten years of continuous monitoring data on any transect subject to an agri-environment scheme. It should be noted that this makes the results less statistically robust than the JBMS full ten-year dataset.

There is good JBMS data from two of the three monitored agri-environmental transects. These both suggest that the schemes were of benefit to butterflies during their time of operation. The other JBMS agri-environment transect was only in operation for two years across one field. Figure 6 shows the annual abundance of butterflies on two JBMS transects: Field A; and Field G (see Table 2).

The planting of bird friendly crops started on Field A in 2007 and has continued through to the present day. The JBMS data from this transect suggests there has been a steep increase in butterfly numbers since the scheme began, so that by 2013 the transect was rivalling some of Jersey's semi-natural habitats in terms of abundance and diversity and the presence of habitat specialist species.

It should be noted that Field A is located close to Les Landes, a prime butterfly site on the north-west of the island. This proximity may have boosted the butterfly population in Field A as it is easy for colonies on Les Landes to spread into the adjacent agri-environment area. Such a spread seems to have started approximately four years after the conservation scheme began operating suggesting that there may be a time lag before significant biodiversity benefits are seen.

Field G was subject to a grassland management scheme between 2006 and 2010. Before and after this time the land was used for intensively farmed crops. JBMS monitoring (which started in 2005) suggests that butterfly numbers were elevated during the operation of the agri-environment scheme and that they have since declined. Unlike Field A, the biodiversity benefits of this scheme were detectable during the first year of operation.

This basic analysis suggests that agri-environment schemes on Jersey have benefited the local butterfly population and could be an effective tool to assist with their conservation. This is in line with results from the UKBMS which suggest that agri-environment schemes benefit local butterfly populations (Fox *et al*, 2006).

Organically Farmed Land

As well as agri-environment schemes, the JBMS also operated across four transects that were, for the entire period of monitoring, certified as organic by the Soil Association.

Certified organic farming requires that crops and animals are raised using natural fertilisers and without the use of pesticides and herbicides. In practice this means that the land can be used for conventional crops such as onions and carrots, and that livestock (including cattle, pigs and chickens) can be reared on fields.

Of the four organically farmed transects monitored by the JBMS, data from one transect was too patchy to be of use (unfortunately, cattle in the fields prevented regular monitoring). The other three transects were suitable and their combined monitoring data are shown in Figure 7.

The trend is variable but shows a moderate decline in butterfly abundance on the organic transects. This trend is perhaps surprising as it might be expected that a reduction in herbicides and pesticides would benefit insects such as butterflies.

Studies outside of Jersey suggest that butterfly populations do benefit from organic production techniques but that this is dependent on the area of organic farming taking place within the wider landscape (e.g. Rundlöf *et al*, 2008). It is therefore conceivable that the monitored organic sites on Jersey were not large enough to benefit the local butterfly population. Additional studies of organically managed land in Jersey are desirable to try and understand the JBMS results better.

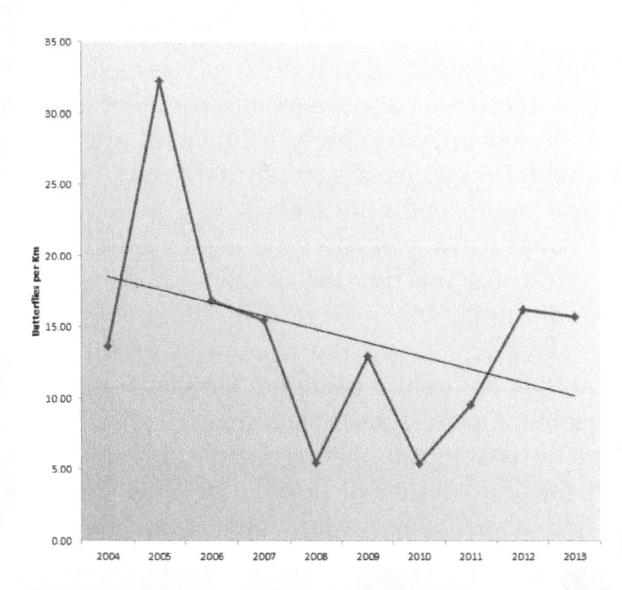

Figure 7. Butterfly abundance (2004 - 2013) on organically farmed transects.

3.4 - Habitat Tolerance

When assessing the health of Britain's butterflies, the UKBMS used the classification scheme of Asher *et al.* (2001) to divide their butterfly species into one of two ecological classes: habitat specialists; and those of the wider countryside.

Habitat specialist butterflies are those species which require particular ecological requirements in order to survive. They are generally limited to a small number of food plants found in a select range of semi-natural habitats and cannot breed in agricultural or urban areas. Wider countryside species are generalists that can tolerate a wider range of habitats including some agricultural and urban ones such as hedgerows, roadside or field verges, gardens and parkland. This distinction is important because when habitats or ecosystems come under pressure, it is usually the habitat specialists that will disappear first, making them a bell-wether for local environmental degradation.

The UKBMS found that Britain's habitat specialist butterfly species are declining at a much faster rate than countryside specialists. This is primarily ascribed to the increased fragmentation and isolation of semi-natural habitats and a loss of natural areas and brown field sites to development and intensive agriculture (Fox *et al.*, 2006).

Jersey's Habitat Specialist Species

Of Jersey's common butterflies, just three are habitat specialists: the Grayling, Green Hairstreak and Swallowtail (see Table 3). These species are all restricted in their distribution with the Grayling and Green Hairstreak being mostly restricted to two JBMS transects (Les Landes, Les Blanches Banques) while the Swallowtail was resident at two, possibly three, sites. The Grayling and Green Hairstreak are doing well on the sites where they occur while the Swallowtail has steeply declined and may no longer be resident in the island.

There have historically been other resident habitat specialist species in Jersey, such as the Large Chequered Skipper, Dark Green Fritillary, Glanville Fritillary and Bath White but all are now locally extinct.

A general lack of habitat specialists and the restricted distribution of the Grayling and Green Hairstreak suggests that the same decline and step-wise extinction of habitat specialist that was observed in UK butterflies has happened here.

Jersey's Grayling and Green Hairstreak colonies are currently surviving on isolated habitat fragments in the west of the island. These populations are increasing but unless they can expand into new areas, the individual colonies will remain vulnerable to unexpected events such as fires.

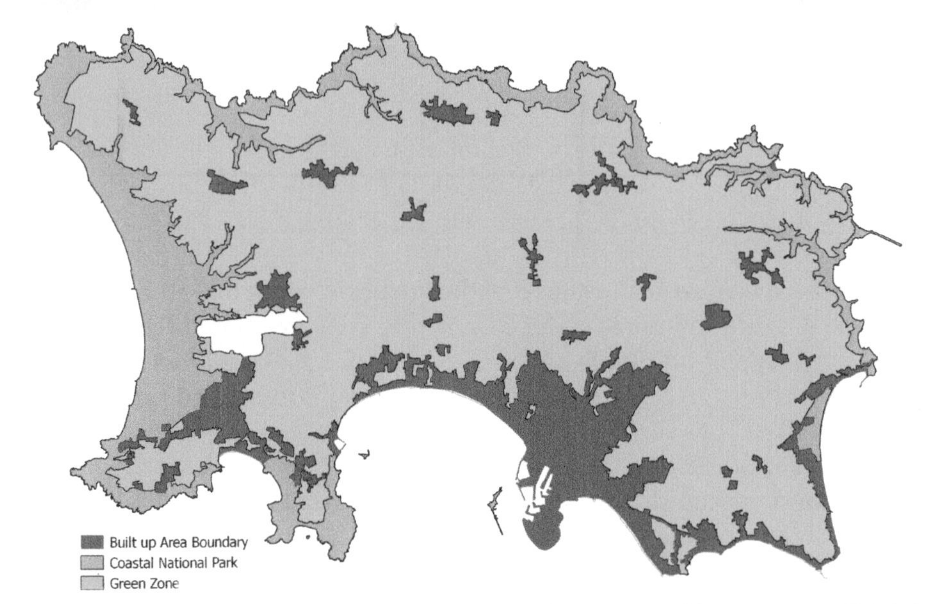

Figure 8. Jersey's principal Island Plan (2011) zones. A majority of the prime semi-natural habitats are in the Coastal National Park. The Green Zone is dominated by agricultural and woodland habitats. The Built Up Area contains concentrated housing and urban habitats.

Jersey's Wider Countryside Species

Jersey's wider countryside species (which includes almost all the common species) are faring better than the habitat specialists but even within this more tolerant group of butterflies there is cause for concern.

An analysis of the commonest butterflies within the three broad land management categories monitored by the JBMS (see Section 3.2) shows that those species found in agricultural and urban habitats have populations that declined between 2004 and 2013 while those associated with semi-natural habitats have generally increased (Table 4).

This suggests that it is only on Jersey's semi-natural sites that most butterfly populations are increasing and that everywhere else they are predominantly decreasing.

For example, just one species (the Small White) is more abundant in agricultural areas than elsewhere in Jersey and even this is showing a steep population decrease. In urban areas the Speckled Wood and Small Tortoiseshell are more abundant than elsewhere but other than these three butterflies, all other JBMS species are more abundant in semi-natural habitats.

Thus while almost all of Jersey's butterflies are wider countryside species, most are faring far better inside the island's semi-natural sites than elsewhere. If this trend continues then in the longer term Jersey may end up with a majority of its butterfly population restricted to a few semi-natural areas while other parts of the island will be dominated by a handful of tolerant or strong-flying species.

Semi-natural	Butterflies per Km	Change 2004-2013
Gatekeeper+	25.15	-22%
Common Blue+	21.20	28%
Meadow Brown+	17.13	90%
Green Hairstreak*+	13.63	458%
Small White	12.61	-67%
Grayling*+	11.06	48%
Small Heath+	10.50	16%
Speckled Wood	10.23	-8%
Large White+	9.03	-21%
Wall Brown+	5.74	97%
Painted Lady+	5.40	-55%
Red Admiral+	3.82	1%

Agricultural	Butterflies per Km	Change 2004-2013
Small White+	9.71	-67%
Speckled Wood	8.18	-8%
Large White	6.22	-21%
Meadow Brown	5.15	90%
Gatekeeper	5.12	-22%
Painted Lady	3.00	-55%

Urban	Butterflies per Km	Change 2004-2013
Speckled Wood+	13.53	-8%
Small Tortoiseshell+	8.07	95%
Small White	7.91	-67%
Large White	6.23	-21%

*Table 4. Lists of the commonest butterfly species for Jersey's main land management categories together with information on their abundance (Butterflies per Km) for that category and island-wide ten-year population trend (Change 2004-2013). * = The species is a habitat specialist. + = The species is more abundant in this category than elsewhere.*

3.5 - Grassland Indicator Species

The European Union (EU) uses butterfly monitoring data from 19 countries to study the trends of 17 grassland-associated butterfly species. The latest results (van Swaay, 2014) suggest that between 1990 and 2011 there has been almost a 50% decline in Europe's grassland butterflies. This decline is ascribed to agricultural intensification in lowland areas and, conversely, the abandonment of grazing and other traditional management techniques in mountain and wetland areas.

Of the 17 grassland indicator butterflies monitored by the EU, Jersey has six of the seven listed 'widespread' species (see Table 5) but none of the habitat specialists. The JBMS has population trend data for five of these species but not for the irregularly recorded Orange-tip.

All five of the JBMS grassland indicator butterflies show a positive change in population ranging between 16% (Small Heath) and 97% (Wall Brown). The average population change for the five Jersey species between 2004 and 2013 is +52% which is in stark contrast to the European wide figure of almost -50% (see Table 5).

However, the apparent success of these grassland indicator butterflies in Jersey needs to be viewed against these species' distribution which is heavily biased towards semi-natural sites (see Section 3.4). In this respect applying the EU's grassland indicator measure to Jersey's butterflies will offer more of an insight into the health of the island's semi-natural grassland sites than its agricultural or urban areas, within which the grassland indicator species are rare or absent.

It is possible that the steep decline in grassland species currently being measured on EU agricultural sites may have already occurred in Jersey prior to the start of the JBMS leaving the island's grassland species mostly confined to semi-natural areas. However, according to the EU criteria, the island's semi-natural sites have been successful at promoting grassland butterfly species.

Butterfly Species	JBMS Change: 2004-2013	EU Trend: 1990 - 2011
Common Blue	+28%	Moderate decline
Meadow Brown	+90%	Moderate decline
Orange-tip	N/A	Stable
Small Copper	+51%	Moderate decline
Small Heath	+16%	Moderate decline
Wall Brown	+97%	Moderate decline
Average	**+52% (5 species)**	**-50% (17 species)**

Table 5. Population changes for the six JBMS species listed as grassland indicators in van Swaay et al. (2013). See also the species entries in Section 2.0.

3.6 - Climate and Jersey's Butterflies

Butterflies are cold blooded animals whose metabolism, behaviour and life cycle are affected by weather and climate. Other studies suggest that butterflies are sensitive to changes in climate and in the UK it has been observed that many southern species are expanding their range northwards as the climate warms (Parmesan, 2003).

Continuous detailed measurement of Jersey's weather began in 1894 and an analysis of the records since then reveals a temperature rise of just over 1°C. This is in line with neighbouring European countries with the island recently experiencing milder winters, warmer summers and the earlier onset of spring weather (Figure 9).

When the JBMS started in 2004 the island's climate had been on a continuous warming trend for two decades but from 2008 onwards north-west Europe experienced a series of cold winters often followed by wet springs and cooler summers.

Rather than rising, after 2008 average annual temperatures fell and rainfall levels increased leading to a series of springs and summers which were less favourable for butterflies. This cooling trend continued until the spring and summer of 2014 which were drier and warmer than the preceding few years (Figure 10).

A comparison of the average annual first sighting for each butterfly species during the first ten years of JBMS monitoring shows a link with the average annual temperature. In warmer years individual butterfly species

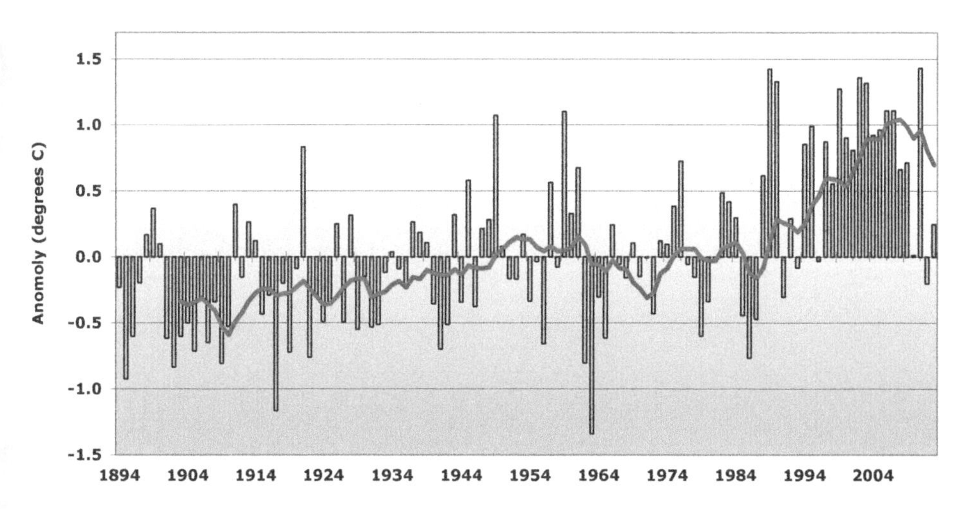

Figure 9. Mean annual temperature on Jersey (1894 to 2013) with the ten-year moving average in red (Source: Jersey Met Office).

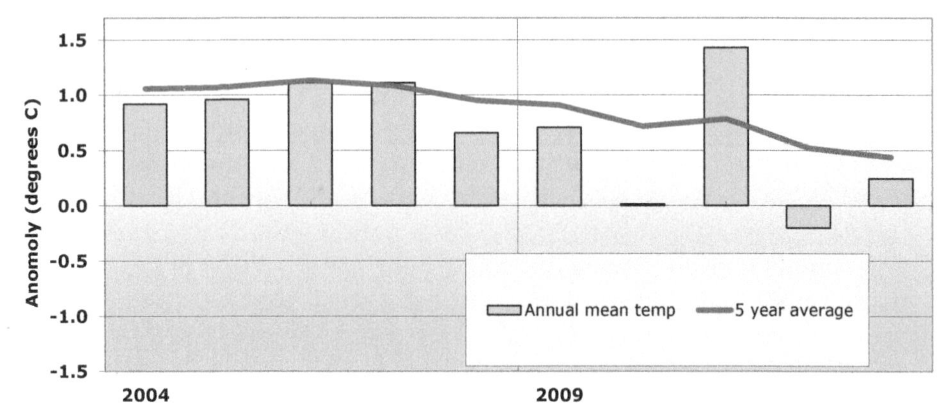

Figure 10. Mean annual temperature on Jersey (2004 to 2013) with the five year moving average (Source: Jersey Met Office).

would be first sighted (on average) earlier than in the cooler years. From 2008 onwards every butterfly reacted to the poorer springs and summers by appearing successively later in each year. This trend could be measured in all of the 24 common JBMS butterflies (e.g. Figure 11).

This suggests that butterflies are efficient indicators of small scale climatic change and that the JBMS data can be used as part of wider phenological monitoring on Jersey.

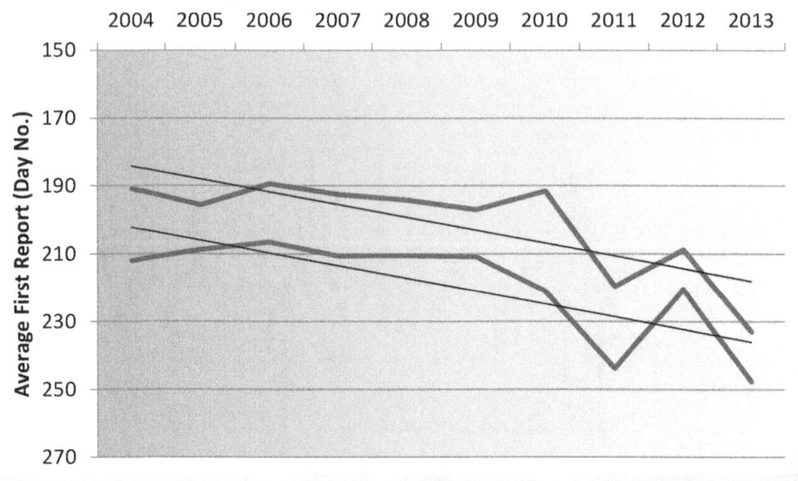

Figure 11. The annual average date (using the day number in the year) of the first reported occurrence of the Essex Skipper (red) and Gatekeeper (blue). The vertical axis scale has been reversed; the higher the day number, the later in the year the butterfly was reported. This suggests that Jersey's butterflies were reacting to the cooling trend outlined in Figure 9.

- Part Four -

Summary and Discussion

4.1 - Summary of Results

The first ten years of the JBMS has produced a dataset which is robust enough to be used to assess the health of Jersey's butterfly populations.

Between 2004 and 2013 volunteer-based monitoring using 38 transects covering a range of habitats and land management styles recorded 34 species of butterfly. Of these 24 may be considered common while a further 10 species are either rare residents or occasional migrants from continental Europe. During the period of monitoring one resident species, the Swallowtail, may have become locally extinct.

A series of analyses were performed on the JBMS dataset in order to discern the state of health of individual butterfly species and to identify any decline or increase within their populations. The results for individual species were compared with those from the UKBMS.

Of Jersey's 24 common butterfly species, ten (41%) show population decreases while 14 (59%) show increases (Figure 12). Over the ten-year JBMS period Jersey's total monitored butterfly population shows a 14% decrease.

When compared with the UK, Jersey's common butterfly species are generally doing better with 18 (75%) showing population trends that are either increasing faster or declining less steeply than in the UK. In the same monitored decade the UK butterfly population declined by 29%.

An analysis of the relationship between JBMS butterflies and the habitats in which they were recorded revealed that the island's agricultural and urban habitats have fewer butterflies in terms of both diversity and abundance. The best of the monitored habitats for butterfly diversity and abundance are all semi-natural.

A majority of Jersey's semi-natural environments are located in coastal areas on the west and north of the island while the centre, south and east of the island contain agricultural, urban or woodland habitats. This has concentrated Jersey's prime butterfly areas into a non-continuous and often thin coastal strip running clockwise from Noirmont headland to Les Landes. Outside of this area the north coast cliffs, Victoria Tower and fringe of Grouville Golf Course are good general areas for butterflies while St Catherine's Woods is a stable habitat for several woodland species rarely recorded elsewhere (Figure 13).

Jersey has just two common habitat specialist species, the Grayling and Green Hairstreak, both of which are restricted to a handful of sites in the west of the island. All other common Jersey butterflies are wider countryside species that can tolerate a greater range of habitats than the specialists. However, the wider countryside butterfly populations are only increasing in semi-natural areas with those in agricultural and urban areas generally decreasing.

An assessment of two monitored agri-environment transects suggests that these conservation schemes are favourable to butterflies but possibly only for the duration of their operation. Organically farmed areas show a decrease in butterfly abundance over time which may be down to their small area within the island's overall agricultural landscape.

The application of the EU's grassland indicator criteria to the JBMS dataset suggests that Jersey's grassland species are doing well in semi-natural areas but declining elsewhere.

The JBMS dataset shows that its data can be matched to measured annual changes in the island's climate. The JBMS data has the potential to be used for phenological monitoring.

The overall conclusion from the first ten-year analysis of the JBMS is that Jersey's butterflies are in an overall decline with the only increasing populations occurring in a handful of semi-natural sites mostly located in the west of the island.

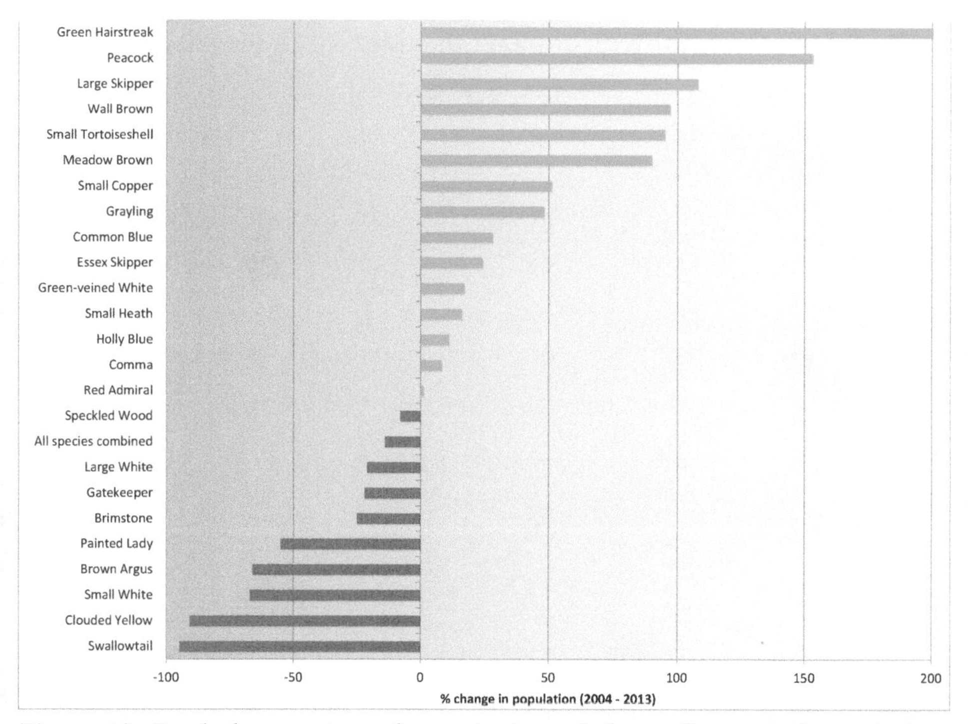

Figure 12. Ranked percentage change in Jersey's butterfly species' populations between 2004 and 2013. Green = positive change in a species population. Red = negative change in a species population. See also Table 3.

4.2 - Discussion

This report has presented the results from a series of analyses into the JBMS dataset. These results paint a mixed picture regarding the overall health of Jersey's butterfly population and suggests that there are issues associated with this that could be explored further.

This section will present some of these perceived issues and will discuss their possible causes and any solutions. The matters raised here are not intended to be presented as a *fait accompli* or as future departmental policy; they are presented as items for discussion.

Habitat Fragmentation

Jersey's high residential population means there is an intense pressure on land usage within the island. Many areas along the southern coastal fringe have been developed for housing while the centre of the island is dominated by agriculture, small housing developments and domestic curtilage.

Most of the island's sizeable areas of semi-natural habitat are situated in the far west, on cliff-tops or along the north coast (essentially the area

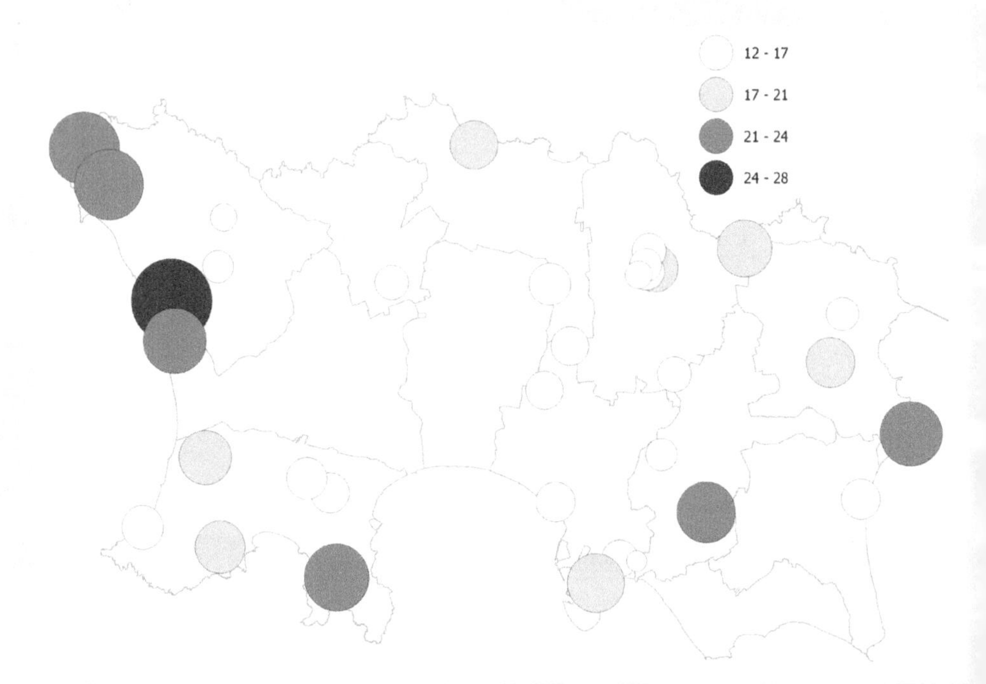

Figure 13. Butterfly abundance (as the average number of butterflies per kilometre walked) at the 38 transects monitored between 2004 and 2013.

covered by the Coastal National Park; Figure 8). Other semi-natural areas, such as woodlands and wetlands, exist in inland valleys or at isolated locations.

The need for housing, amenities and agricultural land has fragmented Jersey's semi-natural areas separating them from one another by fields, roads, parkland and a mix of other developments (Figure 14). The effect of this has been to restrict the island's best butterfly sites to a handful of landscape fragments most of which are unconnected. This is, in the words of a UK specialist who has seen the JBMS data, 'a worst case scenario'.

European studies suggest that butterfly populations are more resilient when they occur in a series of interlinked colonies across a wider landscape. In such circumstances an individual site that is destroyed by fire, extreme weather or another disaster, can be repopulated by butterflies from a nearby colony. However, if sites are fragmented and isolated from one another, then a destroyed butterfly colony cannot be repopulated from nearby and will remain locally extinct. As colonies become forced into fewer and more remote habitat fragments, the chances increase of the species as a whole becoming regionally extinct through step-wise local extinctions.

This concept of individual species utilising wider landscape connectivity is known as 'metapopulation' and it is thought to be a major factor in the ability of butterflies (and other insects) to survive and thrive in local areas (for further information see Fox *et al.* 2006)

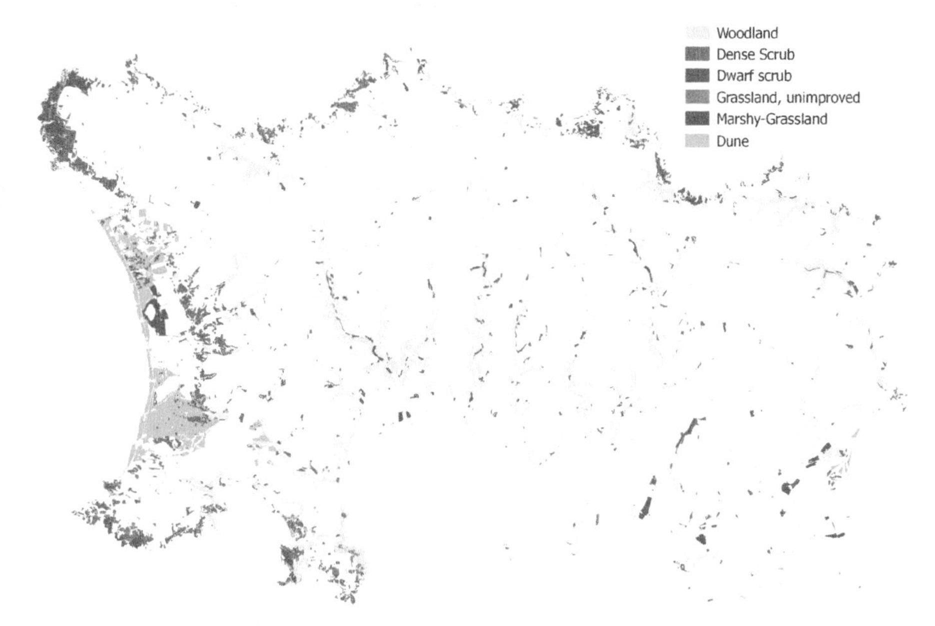

Figure 14. A map of Jersey's principal semi-natural habitats. Note the small gaps within the habitats and the isolation of many sites. Such habitat fragmentation is unfavourable for Jersey's butterfly and general biodiversity.

The effects of habitat fragmentation on butterfly metapopulations has probably been operating in Jersey since the 19th century. The loss of some of Jersey's historical butterfly species such as the Bath White, Glanville Fritillary and Large Tortoiseshell, are probably due to the step-wise extinction of colonies that had become isolated from one another.

The same effect could also threaten some of our current site restricted species such as the Grayling, Green Hairstreak, Common Blue and White Admiral. The development of isolated habitat fragments also makes it difficult for new species to become established as their colonies often have nowhere directly nearby into which they can expand. This may be the case with the Swallowtail which seemed unable to expand away from its isolated resident sites.

Site Restoration and Management

In the UK landscape restoration and management has helped reverse the effects of fragmentation and has even brought back some species from the edge of local extinction (Woodcock *et al.*, 2012). Such initiatives can be costly and time-consuming but the alternative is to risk an ongoing decline in butterfly populations and the local extinction of species.

The JBMS data suggest that, with the exception of Les Blanches Banques where the butterfly population is declining, the island's major semi-natural sites are generally favourable for butterflies. These habitats are the island's best wildlife refuges and to prevent further local extinctions, existing habitat fragments should be expanded through restoration initiatives and, if possible, connected via the creation of wildlife corridors.

In an island the size of Jersey the creation of new areas of semi-natural landscape is problematic but any opportunity to expand or restore habitats within semi-natural sites should be taken. Some locations, such as those with habitat specialist butterflies, may need to be managed with a view to helping selected colonies to remain healthy and expand into new areas.

Positive results from the two monitored agri-environment scheme sites (see Section 3.3) suggest that the tactical expansion of some semi-natural sites could be achieved by implementing conservation measures in adjacent farmland.

Many of Jersey's cliff-top and coastal habitats are fragmented by agricultural land or amenity grassland. Any conservation farming schemes should preferentially be focused on land that is adjacent to these semi-natural habitats or connected through field margins and hedgerows. This is liable to be of greater benefit to the island's wildlife than conservation farming schemes on isolated agricultural sites in the centre of the island.

The predominance of agricultural land in Jersey, especially in the centre and east of the island, is inescapable but management initiatives could be taken which will benefit butterfly and other wildlife species. Such initiatives should be focused on the creation of a wildlife corridor network across the island which, in practical terms, means making efforts to restore hedgerows, managing verges for wildflowers (see Section 4.2) and, where possible, the creation of fallow strips round the edge of fields.

To be successful these initiatives will need to be implemented and then maintained for decades. However, short term measures, such as encouraging the planting of conservation crops, will also benefit butterflies, birds and other wildlife but these should only be considered as stop gap measures. To conserve and enhance our wildlife over the long time will require long term strategies.

The JBMS data also suggest that butterflies are steeply declining within parkland and garden habitats. Although these urban habitats will never rival Jersey's semi-natural sites, they are nonetheless important feeding areas for several species of butterfly including many of the larger, colourful species that please residents and tourists.

Encouraging butterflies into parks and gardens requires planting insect-friendly plants such as buddleia, and lavender and setting aside small areas of tall grasses, nettles and wildflowers These will allow many species of insect, including butterflies, to breed. For example, a handful of gorse plants in West Park (St Helier) is thought to be responsible for regular reports of Green Hairstreak butterflies there. A comprehensive list of insect friendly garden plants is provided on the Royal Horticultural Society's website.

Encouraging contractors, landowners and domestic gardeners to create insect friendly areas is a quick and easy way to help reverse the decline of butterflies and other pollinating species within Jersey's urban and suburban areas.

Food Plants and Countryside Management

The life cycle of butterfly species is intricately linked to the plants on which they lay eggs, feed (as caterpillars and adults) and pupate. Across Europe the post-War destruction of traditional wildflower meadows has been held responsible for declines within many pollinating insect groups, including butterflies (van Swaay, 2014).

Most butterflies have a selected range of plant species on which they will lay their eggs and on which the caterpillars and adults will feed. The presence of suitable plant species is an essential part of a butterfly's life cycle and without them the animal cannot successfully breed.

In recent years there have been campaigns across Europe to recognise the importance that countryside management plays in insect life cycles. Much of this is driven by the need to conserve wildflower populations that exist in hedgerows and verges as these are often remnant habitats from the species-rich flower meadows that existed before the intensification of agriculture (Scottish Natural Heritage, 2013).

The best practice is to cut grassy verges once or twice a year in the autumn or winter (between mid-September and March) after plants have had a chance to set seed. Some areas or stretches of bank facing away from roads should be left uncut for two or three years to prevent loss of structural diversity. All cuttings should be removed to prevent a build-up of soil or nutrients. Hedges and scrub areas need to be managed to prevent them shading out wildflowers but cutting hedge plants every other year is deemed sufficient. Herbicides and pesticides should not be used on verges or hedges (Envision, 2013; Scottish Natural Heritage, 2013).

The value of these practices is recognised at a governmental level and, following an EU ruling, agricultural cross-compliance in England requires a closed period for hedge and verge cutting between 1st March and 31st August (DEFRA, 2014).

Jersey's fields and roadsides are lined with hundreds of miles of grass verge and hedgerow habitats which already play an important role in the life cycle of many animals, including insects such as butterflies. The enhancement of Jersey's hedgerows and verges could provide several key benefits for wildlife including an increased diversity of flowers, pollinating insects, nesting birds and other species. It could help create wildlife corridors across parts of the island which are otherwise species poor.

Moving verge cutting to the autumn or winter, rather than after the harvesting of spring or summer crops, as is often the case at present, would bring significant benefits to Jersey's wildlife. Outside of agriculture similar management practices could be implemented by government and parish authorities on some of Jersey's wider roadside verges and amenity grassland areas as well as by the managers of larger amenity sites such as schools and golf courses.

Although it should be possible to apply wildlife-friendly management practices to many land areas; the management of roadside verges in a similar fashion is more problematic. Prolific vegetative growth along Jersey's roadside verges is a great wildlife asset but it also presents a hazard to motorists, cyclists and pedestrians. Hence the cutting of roadside banks twice a year is a legal requirement under the branchage law.

At present most roadside verge cutting takes place in the two or three weeks prior to the branchage inspections in early July and early September. The need for health and safety is paramount but the cutting of verges and hedges undertaken ahead of the branchage inspections can be unnecessarily severe. The Department of the Environment issues guidance on the best practice for cutting banks, hedgerows and verges (see also Appendix 1).

If this guidance were to be more widely adopted then Jersey's roadside verges, hedgerows and grass banks would present a better habitat for a range of wildlife, including butterflies.

A summary of actions that could be taken to help assist local butterflies in Jersey is provided in Appendix 1. Further advice may be obtained from websites (such as Butterfly Conservation) or the Department of the Environment.

Further Research

The JBMS has proved itself to be an effective means of monitoring biodiversity across a range of habitats and environmental management areas on Jersey. It is expected that JBMS monitoring will continue into the foreseeable future with the next full analysis taking place after its fifteenth or twentieth season of operation.

In the meantime this report has highlighted a number of issues that could benefit from further investigation or closer monitoring. This could include:

- Obtain better data on the health of butterfly populations in woodland, bracken, garden habitats and roadside verges. There is not enough information from these habitats at present.

- Obtaining better monitoring data for conservation farming schemes. To date some of these schemes have been too short-term to provide statistically meaningful results.

- See if the management of areas for wildlife within public sites (such as parks) can encourage butterfly populations.

- See if the specific management of some semi-natural sites for certain butterfly species (such as the Swallowtail) can help halt local declines.

- Investigate Les Blanches Banques to understand why butterflies are declining there and what can be done to reverse this.

Appendix I - A Jersey Butterfly Action Plan

Semi-Natural Areas

- Keep areas of scrub and bracken under control to encourage the development of grasses and wildflowers. This will allow a diverse range of food plants to develop for larvae and adult butterflies.

- Avoid cutting permanent grassland areas between March and September. This is to allow wildflowers to grow, bloom and set seed creating structural diversity within vegetated areas.

- Increase Jersey's current semi-natural area by site expansion or restoration and by connecting together existing habitat fragments. This will increase general biodiversity and help reduce the local species decline associated with habitat fragmentation.

- Manage some areas to assist specific species. Some insect species require very specific plants or habitat conditions in order to breed successfully. Sometimes this can only be achieved through careful management.

Urban Areas

- Set aside small areas within gardens and some public areas, such as parks, cemeteries and coastal strips, for insect-friendly plants. Even small areas of rough grass, nettles and insect-friendly flowers will attract wildlife.

- Ensure that Jersey's major landowners/managers are aware of insect and other wildlife friendly management techniques. It is possible to manage large open amenity areas, such as golf courses, so that they are of benefit to humans and wildlife. Many of these sites already have wildlife management plans and those that haven't should be encouraged to do so.

Agricultural Areas

- Avoid cutting grassy verges and hedges between 1st March and 31st August. Only cut hedge plants every two years. This allows plants to flower and set seed, increasing structural diversity on banks and verges.

- Restore hedgerows with a native plant mix. Hedgerows are vital habitats for a range of plants and animals but many have been neglected and are in a poor state. The restoration of hedgerows using native plants (rather than evergreen and fast-growing shrubs) will provide the maximum benefit for wildlife.

- Encourage conservation farming schemes and techniques, especially on land adjacent or near to existing semi-natural habitats. Conservation farming schemes, such as Jersey's Birds on the Edge project, benefit wildlife in many ways. They are most effective when undertaken near to existing biodiversity hot spots.

- Reduce the use of pesticides, herbicides and fertilisers to encourage animals and flowers. Some farming chemicals will affect useful species as well as pests; their use should be kept to a minimum and targeted.

Hedgerow, Grass Verge and Branchage Law Guidelines

Adopting the best practice guidelines for hedgerow and verge cutting (especially in relation to the Branchage Law) will greatly benefit the island's biodiversity. The following measures are recommended:

- Carry out major hedgerow pruning and branch cutting during the winter months. This will avoid disturbing nesting animals and leave nuts and berries on the trees.

- In the summer months only cut hedges and banks that are affected by the Branchage Law. Do not cut the inner margins of hedges, verges or banks between March and September. This will allow the plants to flower and seed, increasing general biodiversity.

- Manage hedges, banks and verges by trimming leafier overhanging vegetation to a height of 10 cm rather than cutting to the soil level. This will leave roots intact, allow vegetation to recover and prevent soil erosion.

- If possible use hand tools or a strimmer rather than a tractor and flail. Hand tools are more selective and precise than flails and can be used to cut round rare plants or leave small tussocks for insects. If using a flail, try lifting it a few inches from the ground level.

- Never use chemicals to clear vegetation. Aside from potential harm to all wildlife, these can create bare earth patches which will erose and fall into the road.

- Check all hedgerows and banks for nesting animals, rare plants (such as orchids) or invasive species such as Japanese knotweed. If discovered, these areas should be marked off; then please contact the Department of the Environment for further advice.

- Clear all cuttings away from the bank. Either remove them or pile them carefully along the base of the bank. This prevents cuttings from rotting down on the bank itself, creating unwanted nutrients and blocking the growth of new plants.

Bibliography

Ansted, DT and Latham, RG, 1862. *The Channel Islands*, Allen and Co.

Asher, J, Warren, M, Fox, R, Harding, P, Jeffcoate, G and Jeffcoate, S, 2001. *The Millennium Atlas of Butterflies in Britain and Ireland.* Oxford University Press.

Baker, M, 1992. *The Butterflies of Jersey: Monitoring and Rare Species Conservation.* Unpublished MSc Thesis, University College London.

Blacklock, P, 1994. *Butterfly Monitoring Scheme: Jersey 1994.* Unpublished MSc Thesis, University of Hertfordshire.

Brereton, T, 2004. Farming and butterflies in Britain. *The Biologist* 51: 32-36.

Clarke, H, 1991. *The Butterflies of Jersey: A Conservation Report.* Unpublished MSc Thesis, University College London.

DEFRA, 2014. *The Guide to Cross Compliance in England 2014.* DEFRA/Rural Payments Agency.

Envision, 2013. *Road Verge Pilot.* Envision Community Report.

Fox, R, Asher, J, Brereton, T, Roy, D and Warren, M, 2006. *The State of Butterflies in Britain and Ireland.* Pisces Publications.

Halliwell, AC, 1932. The Lepidoptera of Jersey. *Société Jersiaise Annual Bulletin*, 12: 110-117.

Le Quesne, WJ, 1946. The Butterflies of Jersey. *Entomologists' Magazine.* 82: 22-23.

Le Sueur, F, 1976, *A Natural History of Jersey.* Butler and Tanner.

Long, R, 1970. Rhopalocera (Lep.) of the Channel Islands. *Entomologists' Gazette.* 21: 241-251.

Long, R, 2009. A systematic review of the Lepidoptera of Jersey: Part 2: Hesperioidea to Drepanoidea. *Société Jersiaise Annual Bulletin* 30(1):93-99.

Luff, WA, 1908. The Insects of Jersey. *Proceedings and Transactions of the Société Guernesiaise.* 5: 482-511.

Parmesan, C, 2003. Butterflies as bioindicators for climate change effects. In *Butterflies – Ecology and Evolution Taking Flight* (ed. Boggs, CL *et al.*), University of Chicago Press, pp.541-560.

Picquet, FG, 1873. A list of butterflies inhabiting Jersey. *The Entomologist.* 6: 399-401.

Pollard, E, Yates, TJ, 1993. *Monitoring butterflies for Ecology and Conservation.* Chapman and Hall.

Rickell, R, 1990. *Butterfly Monitoring Project at Petit Pré, Trinity.* Unpublished BSc Dissertation, Hatfield Polytechnic.

Riley, ND, 1922. Jersey Lepidoptera. *The Entomologist.* 58: 149-151.

Rundlöf, M, Bengtsson, J and Smith, HG, 2008. Local and landscape effects of organic farming on butterfly species richness and abundance. *Journal of Applied Ecology*, 45(3): 813–820.

Scottish Natural Heritage, 2013. *The Management of Roadside Verges for Biodiversity*. Commissioned Report No. 551.

Shaffer, M, 2008. *Channel Islands Lepidoptera*. Privately published.

States of Jersey, 2004. *An environmental monitoring strategy for Jersey*. Unpublished report, Department of the Environment (SoJ).

States of Jersey, 2005. *The State of Jersey Report: 2005-2010*. States of Jersey.

States of Jersey, 2011. *The State of Jersey: A report on the condition of Jersey's environment*. States of Jersey.

Thomas, JA, 2005. Monitoring change in the abundance and distribution of insects using butterflies and other indicator groups. *Philosophical Transactions of the Royal Society B* 369: 339-357.

van Swaay, CAM, 2014. *The European Grassland Butterfly Indicator: 1990–2011*. EEA Technical Report No. 11/2013.

van Swaay, CAM, Nowicki, P, Settele, J and van Strien, AJ, 2008. Butterfly monitoring in Europe: methods, applications and perspectives. *Biodiversity and Conservation*, 17(14): 3455–3469.

Woodcock, BA, Bullock, JM, Mortimer, SR, Brereton, T, Redhead, JW, Thomas, JA, Pywell, RF, 2012. Identifying time lags in the restoration of grassland butterflies communities: a multi-site assessment. *Biological Conservation* 155: 50-58.

Woods, R, 2007. *A study of the Spatial Population Structure of the Grayling (*Hipparchia semele*) at Les Landes SSI in Jersey*. Unpublished Report, States of Jersey.

Identification Guides and Websites

Anon., 2010. *Guide Atlas des Papillons des Côtes d'Armor*. VivArmor Nature.

Bebbington, J, 1998. *Guide to the Butterflies of Britain*. Field Studies Council: Identification Sheet.

Lewington, R, 2003. *Pocket Guide to the Butterflies of Great Britain and Europe*. British Wildlife Publishing.

Thomas, J, 2014. *Philip's Guide to Butterflies of Britain and Ireland*. Philips.

Tolman, T, Lewington, R, 2009. *Collins Butterfly Guide*. HarperCollins.

Butterfly Conservation: www.butterfly-conservation.org

States of Jersey: www.gov.je/Environment

UK Butterfly Monitoring Scheme: www.ukbms.org

Acknowledgements

The Department of the Environment thanks the following people, especially those who have walked our transects over the years, for their help and assistance with the Jersey Butterfly Monitoring Scheme and the production of this book:

Nick Aubin, Rich Austin, Margaret Austin, Joan Banks, John Banks, Irene Beaumont, Tony Beaumont, Benjamin Bisson, Marc Botham, Bertram Brée, Tom Brereton, Paul Chambers, Nick Channing, Sue Clarke, Nina Cornish, Liz Corry, Dianne Custance, Charles David, Donna De Gruchy, Peter Double, Ian Everson, Mike Freeman, Richard Fox, Tizane Gallichan, Caroline Germain, Alan Gicquel, Tanya Gicquel-Tirel, Henry Glynn, Gordon Hall, Sue Hardy, Ian Harrison, Richard Huelin, Jon Horn, John Jackson, Viviane Jayes, Sarah Kitchin, Michael Krigu, Caroline Leach, Pauline Le Bailey, Zoe Le Cornu, Tim Liddiard, Margaret Long, Roger Long, Sarah Maguire, Denise McGowan, Carol Maindonald, Shelia Mallet, Christian Marcos, Jean Marett, Christiane Mccarthy, Julia Meldrum, Jo Moss, Lindsey Napton, Richard Perchard, John Pinel, Annie Pope, David Pope, Judith Pountney, Chris Quérée, Tim Ransom, John Rault, Mike Romeril, David Roy, Cristina Sellares, Claire Stanley, Mike Stentiford, Bob Tompkins, Jill Tompkins, David Tipping, Chris van Swaay, Bunty Wagstaffe, Paul Wagstaffe, Pat Walker, David Wedd, Nadine Wohl, Marc Woodall, Tim Wright.

We also thank the following organisations for their help with establishing, running and publicising the Jersey Butterfly Monitoring Scheme and for their assistance with the production of this report:

Butterfly Conservation, Guernsey Biological Records Centre, Jersey Biodiversity Centre, National Trust for Jersey, Société Guernsiaise, Société Jersiaise, States of Jersey (Department of the Environment), UK Butterfly Monitoring Scheme and all the landowners and site managers that have allowed us to establish transects on their properties.

Picture Credits

Below is a list of the photographers who kindly agreed to donate their images to this book. We are very grateful to them for allowing us to use their work. The initials in brackets may be used to identify the images on the pages listed below.

Richard Perchard (RP): All Cover Images. Pages: 26, 27(x2), 28, 29, 30 (x2), 31 (x2), 32 (x2), 33 (x2), 34 (x2), 35, 36 (x2), 37, 38, 40 (x2), 41, 42, 43 (x2), 44 (x3), 45, 46 (x2), 47 (x2), 48 (x4), 49, 50, 51 (x2), 52 (x2), 53 (x2), 54

Chris van Swaay/Dutch Conservation (CVS): 39, 54 (x2), 55 (x2), 56 (x3), 57, 58 (x2), 59 (x3), 60

Peter Double (PD): Pages: 28,33

Shutterstock (CCL): Pages: 26, 30, 37, 38, 39, 41, 45, 47, 56, 58 (x2), 59, 60, 62 (x2)

States of Jersey: Pages 17, 19, 85 (x2)

Denise McGowan: Page 85

Paul Chambers (PC): Pages 32, 41, 42, 49, 50

Lightning Source UK Ltd.
Milton Keynes UK
UKOW07f0657191115

263064UK00010B/44/P

9 780901 897558